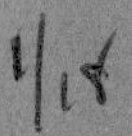

P9-CKF-510

DISCOVERING THE MAMMOTH

DISCOVERING THE MAMMOTH

A TALE OF GIANTS, UNICORNS, IVORY, AND THE BIRTH OF A NEW SCIENCE

JOHN J. McKAY

PEGASUS BOOKS
NEW YORK LONDON

DISCOVERING THE MAMMOTH

Pegasus Books Ltd.
148 W 37th Street, 13th Floor
New York, NY 10018

First Pegasus Books cloth edition August 2017

Interior design by Maria Fernandez

Library of Congress Cataloging-in-Publication Data is available.

ISBN: 978-1-68177-424-4

10 9 8 7 6 5 4 3 2 1

Printed in the United States of America
Distributed by W. W. Norton & Company

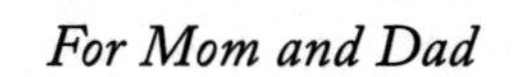

For Mom and Dad

CONTENTS

INTRODUCTION

> "Palaeontology may be said to have been founded on the Mammoth."
>
> —Henry Neville Hutchinson, *Extinct Monsters*, 1896

Charles Darwin spent the last part of September 1833 traveling overland from Buenos Aires to Santa Fe in northern Argentina. As the naturalist on the *HMS Beagle*, it was his job to examine the local geology and collect samples of the plant and animal life whenever the ship landed. He also hunted for fossils. Four days into the trip, his caravan stopped on the banks of the Rio Tercero.

> I staid here the greater part of the day, searching for fossil bones. . . . Hearing also of the remains of one of the old giants, which a man told me he had seen on the banks of the Parana, I procured a canoe, and proceeded to the place. Two groups

> of immense bones projected in bold relief from the perpendicular cliff. They were, however, so completely decayed, that I could only bring away small fragments of one of the great molar-teeth; but these were sufficient to show that the remains belonged to a species of Mastodon.

Darwin mentions mammoth and mastodon bones several times in his memoir of the *Beagle*'s five-year voyage around the world. In these references, we can see him struggling with the problems that would lead him to formulate his theory of natural selection as the driver of evolution. He lists more than a dozen large mammals that had disappeared from the South American landscape, leaving it impoverished and transformed. With the exception of some horse teeth he found, he believed that the large mammals that produced the bones were extinct and that none of them still lurked in some hidden corner of the earth. From the depth and position in the earth where he found the bones, he determined that all of these missing species had lived together and that they had disappeared only recently. He had no doubt that they had all been native to regions where their bones were found. Though he had studied for the clergy, he never considered that their bones had been brought there by the biblical Deluge. He believed that the climate there had once been different (though not substantially so) and that such changes led one collection of lifeforms to be replaced by another. Finally, he believed that those changes worked slowly over many thousands, even millions, of years. Two generations earlier, when his grandfather Erasmus Darwin had formulated his own theory of evolution, all of these ideas had been controversial. A few generations before that, they were unthinkable ideas or, worse, sacrilegious.

In the 1690s, less than two centuries before Darwin roamed the coast of Argentina, literate Westerners became aware of a new type of ivory found in Eastern Russia. Muscovite merchants said the ivory came from an unknown Siberian beast that the natives called "mamant." Their descriptions of this beast ranged from sea monsters to cave-dwelling shape-changers to the biblical Behemoth. The beasts were known only from their remains; no one had ever seen one alive. Westerners realized that the shape of the ivory was similar to that of elephants' tusks, but knew it was impossible for elephants to live in the Arctic. As they struggled to make sense of this information from Siberia, other elephant-like bones were discovered in North America. Older giants' bones and saints' relics from places like Ireland and Germany were reexamined and also recognized to be elephant-like. It began to look as if elephants had once roamed all parts of the earth. How was this possible? Had the biblical Deluge, or a similar cataclysm, transported elephants' bones all over the globe? Had the whole earth once been tropical and home to elephants? Or was the solution even stranger?

At the time these same people became curious about the mysterious "mamant," or "mammoth," the study of fossils was dominated by seashells found in the wrong places, whether it was deep underground or on mountain tops. Yet because the sea is large and its depths unknown, differences in shape from known species—even completely unknown species—were not difficult for people to explain away. The remains of unrecognizable land animals, especially large ones, were a tougher problem. The mysterious mammoth pushed fossil studies in a new direction. Unraveling that mystery required the development of a new, specialized intellectual toolkit. Unthinkable ideas such as extinction and a history of the earth itself separate from, and older than, human history needed to be embraced. Revolutions in geology, comparative anatomy, and taxonomy had to come about. Even folklore was enlisted to shed light on strange bones in the earth. Each advance, being applied to the mammoth problem, provided a template for studying other mysterious remains.

Is it excessive to say that without the mammoth there would have been no paleontology and no dinosaurs? Perhaps. But, without the mammoth as a focusing problem and a catalyst that drove a revolution in thinking, vertebrate paleontology would have taken longer—perhaps much longer—to develop. It's difficult to say when attention would have shifted from seashells to bones. Mastodon bones from North America weren't examined until the 1750s. The first completely unrecognizable vertebrate discovered in Europe was a moasaur found at Maastrict in the Netherlands in 1764 and it was another sea creature, not a land animal. Without knowledge of the Siberian mammoth with its strange name and extreme location, the bones of large land mammals found in Western Europe were easily explained by citing Hannibal's elephants or the Roman circuses. It was the mammoth that forced European thinkers to reconsider giant bones in their own collections. The first frozen mammoth to be recovered was spotted near the Lena delta in 1799. By then, the intellectual toolkit of paleontology had more than a century to be assembled. Many of the basic concepts had been debated and rough consensuses had been achieved. The Western intellectual elite was ready to accept that the world was a very old place and that the mammoth was a lost species that had lived in a place where similar modern elephants could not survive. The past was stranger than they had imagined. It was a liberating moment. The first important paper on the subject listed three extinct species. Within a few years the number had risen to twenty and has continued to rise ever since.

The same year that the Lena mammoth was discovered, Mary Anning was born in Lyme Regis, Dorset, England. Her father, Richard Anning, was a cabinet maker who supplemented his income by beach-combing for fossils eroded out of the nearby cliffs. He sold these to visitors as novelties. Richard Anning died in 1810, leaving the family with no income other than the fossil collecting that Mary and her brother Joseph conducted. Mary and Joseph were very skilled at recognizing and cleaning fossils, at gathering fragments of fossils, and at reassembling them into more profitable wholes. If the fossils they collected had remained curios, the income from them would not have enough to keep the family together. Fortunately, the market had changed since the days when Richard Anning first collected them. The gentlemen who came down from London were no longer dilettante collectors looking for decorative pieces; many were involved in serious scientific pursuits. When Mary and her brother, barely in their teens, assembled the first complete ichthyosaur, no one questioned that such a strange creature had hunted in seas that once covered Dorset. The discovery was written up in the local paper. The fossil was purchased and eventually became the subject of six papers in the journal of the Royal Society (written, of course, by a man of a better class). The easy acceptance of Mary Anning's strange fossils was the end result of a century of studying the mammoth. It was a marked contrast to a time when they would have been viewed as holy relics or "jokes of nature," rocks that only coincidentally resembled bones.

Who discovered the mammoth? Don't answer that; it's a trick question. No one discovered the mammoth for the simple reason that mammoths were never unknown to us. From Cape Town to North Cape, from Spain to Siberia, from Ireland to Indonesia, and from Alaska to Argentina, we have lived among the bones and remains of mammoths and other extinct elephants for as long as we have been human. Long after the last mammoth died and was taken off the menu, our northern ancestors continued to use mammoth products. Their bones were used as building materials and their ivory was used to make tools and art, and as a trade commodity. But there came a time when our ancestors no longer knew what kind of creature the mammoth had been. Each culture interpreted the remains of mammoths and other giants through the lens of their own world view and mythology. When the Classical Greeks saw deposits of giant fossils, they knew they had discovered the battlefields where the gods had vanquished the Titans. When the Chinese discovered buried ivory, they knew they had found dragons' teeth and used them for medicine. When Native Americans along the Ohio River found full skeletons in salt springs, they knew they were seeing the remains of the grandfathers of modern animals. When Northern Siberians found bloody carcasses eroding out of river banks, they knew they had found the recently deceased remains of giant mole-like creatures that caused the frozen ground to heave up in the winter and sink down in the spring. If no one discovered the mammoth, perhaps the question we should be asking is: how did the mammoth once again become a mammoth?

It began with someone finding some bones . . .

CHAPTER 1

GIANTS AND UNICORNS

Early on the morning of Friday, January 11, 1613, a group of workmen, digging in a sand pit near the Castle of Marquis Nicolas de Langon in the Dauphiné province of southern France, happened across the bones of a giant. We don't know the names of the laborers, why they were digging, or what they thought of the bones, but we do know what they did next: they called off the work and sent someone to notify the master of the castle. Their response was more than a simple matter of taking advantage of an opportunity to get out of the weather, though that must have been a consideration; four months earlier, the powerful governor of the province, François de Bonne, the Marshal Lesdiguieres, had written to Langon specifically asking him to be on the lookout for large bones as he would like to have some for his "cabinet of curiosities." These collections of natural history objects and curiosities had become popular in Renaissance Italy and no powerful or fashionable gentleman would dream of being without one. The discovery of large, ancient bones was not unknown in this part of France; the field where the laborers had been digging was known as the Field of Giants and Lesdiguieres made reference to other giant's bones previously found on Langon's lands. The marquis must have been pleased to be able to respond to the marshal so quickly. Before notifying Lesdiguieres, Langon sent for Pierre Mazurier (in some sources his name is given as Mazuyer), the barber-surgeon in the nearby town of Beaurepaire, to examine the bones and confirm the discovery. Mazurier arrived late that same day and confidently pronounced the bones to be those of a giant.

What happened next would be a source of controversy over the next months and years. As the workmen tried to lift the bones out of the pit, many of them fell to pieces leaving only unrecognizable fragments. Naturally, the workers were blamed for mishandling the bones. The accusation was a bum rap. It would have been difficult to save most of the bones. Ancient bones that have not petrified are very fragile things. Collagen rots and acidic water carries away many of the minerals. This is the beginning of petrification. Over a period of time, that can last from hundreds to hundreds of thousands of years, these materials are replaced by dissolved silica from the surrounding earth and compressed to become rock. It is also one of the most vulnerable times for bones on that journey. As exposed bones dry out, deprived of the surrounding soil that had maintained their shape for so long, they become brittle and delicate. Only the densest parts of bones survive very long out of the ground without careful preparation. The old horror movie trope of a long-hidden skeleton turning to dust at the first touch has a basis in reality. When Mazurier arrived to look at the bones, they had already been sitting in the open air for ten hours. Frustratingly, the skull was among those that disintegrated. In many species, a skull might look solid but, in reality, it's nothing more than a series of thin plates honeycombed by sinuses. This is especially true of elephants and their relatives. Later, when the identity of the bones was being debated, the convenient lack of a skull would be called out as proof of fraud. Of the bones Mazurier recovered, the best surviving pieces were two sections of jaw, with teeth, a complete tibia, two vertebrae, a rib, some ankle and foot bones, and the end parts of some of the long bones. All of the surviving bones were from the left side of the skeleton, which was deeper in the ground and partially petrified. Mixed in with the bones were some silver coins or medallions.

When he heard the news of the discovery, Lesdiguieres had Langon send some of the bones to the bishop of Grenoble, who gave them to the doctors at the university in his town to identify. The good doctors agreed with Mazurier and proclaimed them to be the bones of a giant. It's safe to assume that these were the very best bones and that at least one of them ended up in Lesdiguieres collection. Lacking a skull, part of a nice femur or some teeth always make good additions to a cabinet of curiosities. Lesdiguieres collection has not survived so we can't know for sure what he chose. At the same time, Langon sent some of the other bones to the university in Montpellier, home to a medical school. The doctors there, possessing state-of-the-art knowledge of anatomy, also pronounced them to be the bones of a giant. Confident now that he had a box of genuine giant's bones, the question for Langon was what to do with them. Mazurier had an idea. He was sure people would pay to see the bones of a giant and convinced Langon to let him tour the country with them. Mazurier drew up a contract giving him exclusive rights to show the bones for eighteen months unless the king decided to purchase them. By late March, he was on the road to Paris.

Bones of giants would indeed be impressive, but many churches in France in the late seventeenth century already had a bone or two from a giant. What made Mazurier so confident that people would pay to see Langon's bones was that these bones came with a good story. Mazurier commissioned a Jesuit from Lyon, named Jacques Tissot, to write a pamphlet. The title tells the whole story:

> True history of the life, death, and bones of Giant Theutobochus, King of Teutons, Cimbri and Ambrones, defeated 105 years before the coming of our Lord Jesus Christ. With his army, which numbered four hundred thousand combatants, he was defeated by Marius, the Roman consul, killed and buried near the castle called Chaumont, and now Langon, near the town of Romans in Dauphiné. There his tomb was found, thirty feet in length, on which his name was written in Roman

> letters, and the bones therein exceeded 25 feet in length, with one of the teeth weighing eleven pounds, as you will see the in the city, all being monstrous in both height and size.

Who was this Theutobochus and how did he become identified with the bones? As the title of the pamphlet says, he was a barbarian king who threatened Rome at the end of the second century BCE. Not much is known about Theutobochus except that he was very large. The early Christian historian Paulus Orosius wrote that Theutobochus could "vault over four or even six horses" and that he towered above other men. Theutobochus was one of the leaders of a confederation of tribes that had been displaced from Denmark or northern Germany around 115 BCE. According to the Roman historian Florus, their lands were made uninhabitable by "inundations of the sea." The Roman historian and geographer Strabo expressed doubt that such a thing was possible, but the phrase could easily describe a storm surge similar to that which flooded New Jersey and New York during Hurricane Sandy in 2012. Coastline-changing storms of this sort hit the North Sea every few generations. The Grote Mandrenke (Great Drowning of Men) of 1362 killed more than 25,000 people in the same region that the tribes under Theutobochus had called home and the second Grote Mandrenke in 1634 destroyed the Island of Strand. During the three centuries before the tribes began their trek, the North Sea had risen almost two meters, which would have made lowlands very vulnerable to inundation. After wandering through Central Europe for almost a decade, the tribes set their eyes on the fertile lands of the Po valley in northern Italy. The Roman Senate refused to allow them to enter Roman lands and sent an army to stop them from entering the valley. The northerners destroyed it. The Senate sent a second army and the northerners destroyed that one, too. They sent a third army and Celtic tribes in what's now Switzerland destroyed it before it even reached the northerners. The fourth and fifth armies were sent together and, while the commanders squabbled over who was in charge, the northerners and the Celts joined forces to destroy them separately. At this point, rather than march on Rome, the tribes split up, with one part moving northwest to plunder Gaul (France) and the other southwest to plunder Iberia (Spain). During this lucky respite, Gaius Marius returned to Rome.

Marius was the ablest general of his generation. He had just concluded a victorious war in North Africa and been elected consul, the highest office in the Roman Republic. Through reforms enacted during an earlier term as consul, he was popular with the troops and the lower classes. This broad base made him by far the most powerful man in Rome. Whatever fears the senators may have had about concentrating too much power in the hands of one man, they feared the sack of the city and the destruction of their country even more. As preparations dragged on for the campaign against the German and Celtic invaders, they elected him to an unprecedented—and illegal—second, consecutive consulship. Marius gathered a force of veterans from Africa, bolstered by new conscripts, and entered Provence, an area that been annexed to the Roman Republic just fourteen years earlier. Two years passed, with Marius being elected to two more consulships, before the northerners returned to resume their march on Rome. Marius used those years training his new army and pacifying local Celtic tribes so that the northern barbarians would find no new allies and his armies would be hardened and confident after actual victorious combat.

In the summer of 102 BCE, Theutobochus led portions of the migrating tribes, made up of the Teutones, Ambones, part of the Cimbri, and one the revolting Swiss tribes, toward the route that Hannibal had used to cross the Alps 120 years earlier. This route led from the Rhône River, up the valley of the Isere, and over one of several possible passes to enter the Po valley from the northwest. Marius anticipated this route and had placed his army in a well-fortified camp at the junction of the Rhône and Isere Rivers, not far from where the Marquis Langon would have his lands seventeen centuries later. Theutobochus led his warriors in unsuccessful attacks against Marius's defenses for three consecutive days. On the fourth day, rather than continue the attacks, Theutobochus broke off and led his people south hoping to take the coastal road along the Riviera to invade the Po valley from the Southwest. Marius waited until the entire horde was on the road, then broke camp and followed them. The Romans attacked and beat the barbarian rear guard and, full of confidence, raced past the main body of the horde to build a new fortified camp on high ground across the coastal road at the Roman settlement of Aquae Sextiae near the modern city of Marseilles. Once again Marius let Theutobochus attack his prepared position, this time uphill, under the unforgiving Mediterranean sun. Late in the day, the Romans counter-attacked, routing the northerners and killing tens of thousands. The fate of Theutobochus's is unclear. Orosius says he was killed in the battle. Florus says he was taken alive to Rome for Marius's triumphal parade, where, "being a man of extraordinary stature, he towered above the trophies of his defeat." According to Roman tradition, he would have been executed in Rome immediately afterward.

How Theutobochus became identified with Marquis Langon's giant's bones is a bit of an historical mystery that has never been solved. Theutobochus's story was fairly well known in France at the time. Both Orosius and Florus had been translated into common French several times since the advent of printing in the fifteenth century. Ten days after the discovery of the bones, Mazurier made a notarized statement about the discovery to Guillaume Assalin, the local constable. Mazurier gave the constable a very detailed description of the bones, including the skull which he had measured in place before the unsuccessful attempt to lift it out of the pit. Along with this description, Mazurier added two details. He claimed the workers found the bones inside a brick sepulcher thirty feet long, twelve feet wide, and eight feet high. On the front of it, he claimed, was a gray stone engraved with the words "Theutobochus Rex." No trace of the tomb or the stone has ever turned up. He also described the silver medals saying they had the image of Marius on one side and the letters MA on the other.

In the four hundred years since, historians have generally regarded Mazurier as a charlatan who invented the Theutobochus story from whole cloth. If there was a fraud, Mazurier wasn't alone in committing it. It's highly unlikely that, of Marquis Langon, constable Assalin, and the notary, none of them was ever curious enough to look into the hole in the ground and confirm what Mazurier claimed. They would have known whether or not the tomb Mazurier described, or something like it, really existed. When critics began to attack Mazurier, all of these men were silent or unavailable for comment. Though none of them stepped forward to support him, none of them stepped forward to denounce him either. Although Theutobochus's last battle took place at Aix near Marseilles, there was a local tradition in

Dauphiné that battle took place on the stream that passes below the Castle Langon near where the earlier battles between Marius and Theutobochus had happened. The French paleontologist Léonard Ginsburg pointed out in the 1980s that the soil where the workers dug has a brick-like color and often breaks in straight lines. He believes that we should give Mazurier the benefit of the doubt and allow that his imagination caused him to see bricks and Roman letters where there were none and that his enthusiasm was great enough to convince the others to see the same.

We likely will never know the full truth of the discovery. What we do know is that the bones were real. For that, we should thank Mazurier rather than condemn him. The usual fate of giants' bones in Medieval and Renaissance Europe was for them to be put on display at the local church or town hall as evidence of God's majesty and to be brought out for special occasions until they fell to pieces. Alternatively, they might be picked up by a wealthy collector with an interest in the new natural philosophy, such as Marshal Lesdiguieres, and displayed in his cabinet of curiosities, also until they fell to pieces. Had either of these been the case, Langon's bones would have vanished from history. Thanks to Mazurier's showmanship, they did not. Mazurier brought the bones to the attention of a literate audience, who examined, argued over, and wrote about them. What the learned men of France had to say about the bones gives us an excellent view into how they viewed the whole concept of fossils at the beginning of the Scientific Revolution.

On June 18, 1613, Mazurier arrived in Paris with the bones. He set up a tent with a large sign featuring a drawing of the bones and charged the curious a small fee to see the real thing. The show was a huge success. In two months, Tissot's pamphlet ran through three more printings to meet popular demand. The show eventually attracted the attention of the court, as anticipated, and Mazurier received notice that the king would like to have the bones sent to Fontainebleau Palace so that he could look at them. King Louis XIII was twelve years old at the time. What twelve-year-old with unlimited power would not have "asked" to have them sent over? The bones were laid out in the chambers of the queen mother and regent, Marie di Medici. On seeing them, Louis asked a courtier if such giants had ever really existed. Yes, the courtier replied, imagine what a great army such men would make. The king was not enthusiastic about the idea; they would soon eat the country clean, he commented. By all appearances, the court enjoyed the exhibition. The king gave Mazurier a reward for the show, but even he had doubts about the Theutobochus part of the story. Two weeks after the viewing, the king's secretary wrote to Langon requesting the rest of the bones and some parts of the tomb, including the gray stone inscribed with Theutobochus's name, be sent to the court. Four days after that, Mazurier packed up, slipped out of town, and took the bones on a tour of northern France, England, and Flanders. He was last seen taking the bones to Germany. At least one other request for evidence of the tomb was sent from the court to Langon. Langon was away on business when both letters arrived. He stayed away for a year and never responded to the requests for additional evidence. Eventually, the king lost interest and the requests ceased.

Even though Mazurier and the bones had left Paris, people outside the court continued to talk about them for years and debate whether they were really the bones of a giant, and if it was a giant, whether that giant was Theutobochus, and whether they were really bones at all and not natural productions that just happened to look like bones. In late October, a pamphlet entitled *Gygantosteologie, ou Discours des os d'un Géant* (Gygantosteologie, or Speech on the Bones of a Giant) appeared in Paris. The author was Nicolas Habicot, a member of the barber-surgeon's guild, the same guild as Mazurier. Habicot defended Mazurier's claim that the bones were those of a human giant and specifically those of Theutobochus. Habicot's pamphlet had three parts. First, he repeated Tissot's story. Second, he explained how his own medical examination of the bones convinced him that these were the real remains of a human giant. Third, he made a more general argument for the historical existence of giants. Habicot was able to add many facts about the discovery that were not in Tissot's pamphlet and that he could only have learned by talking to Mazurier, most importantly the measurements of the lost skull. Responding to the suggestion that bones might be those of some large animal such as a whale or elephant, Habicot explains that this is not possible. Man, he writes, possesses a soul as well as other unique attributes such as the bones in our hands that allow us make all manner of useful tools and the heels on our feet that allow us to walk upright. That last was very important to him. Two of the most identifiable bones Mazurier brought to Paris were a heel and and ankle bone. He admits he has never seen an elephant, but he knows it to be true that they do not possess these traits. Moving from the specific to the general, he delivers a passionate argument for the historical existence of giants. In this, he brings together examples from the Bible, classical mythology, history, chivalric poetry, and modern rumors, ultimately building an argument that amounts to little more than "so many eminent writers couldn't all be wrong, could they?"

Habicot and the king's courtier were not out of the mainstream in believing in the historical existence of whole nations of giants as opposed to occasional men of unusual height. The belief that giants had once walked the earth and tormented our ancestors was once found in mythologies all over the world. I won't go so far as to say the belief was universal—anthropologists hate it when you say anything is universal—but it was very common. At the beginning of the seventeenth century, the literate elite of Europe was divided over the question of giants. Those challenging the idea had to go against traditions that pre-dated Christianity. In the thirteenth century BCE all the major Mediterranean civilizations underwent a simultaneous collapse. Historians argue passionately about the causes of this collapse, the sequence of events, and how bad it was in each region. Some areas, such as Greece, were plunged into a dark age that lasted almost five hundred years. When they returned to an urban, literate society, a great deal of their specific knowledge of the past had been lost, including their writing system. Surviving records were indecipherable. A new writing system had to be borrowed from the Phoenicians and modified to the Greek language. Events like the Trojan War had become legendary narratives in which gods and demigods participated beside mortal men and women. To Classical Greeks, the ruins of pre-collapse temples, fortresses, and palaces seemed impossibly large, the sort of thing that only men of gigantic stature could have built. This idea fit in well with the idea that mankind and the world in general were in decline. Hesiod, writing around

700 BCE, left the earliest, surviving, Greek account of the Ages of Man. From the nearly perfect world of the Golden Age, Hesiod wrote that man declined into the Silver Age, the Bronze Age, the Heroic Age (a slight improvement), and into his own age, the Iron Age. In each age, mankind had a shorter lifespan, worse health, smaller stature, and more strife as the world itself became worn out. Eventually, mankind will reach the Clay Age and everything will fizzle out altogether. Six hundred years later, the Roman poet Lucretius looked at the world and lamented "what we are witnessing is already in its decay: the Earth has lost the creative power of the past, animals no longer produce the gigantic size of those early days, the ground is no longer capable of spontaneous fertility."

Fossils contributed this dismal world view. The Classical Greek world surrounding the Aegean Sea has a great number of bone beds filled with the remains of large mammals, animals much larger than any known to the Greeks. A large percentage of these bones are from mammoths and other proboscideans. By interpreting these bones as human, the Greeks found one more proof of the idea of eternal decay. Various bone beds became identified as the specific locations of famous battles between the gods and the Titans. The Greeks and later the Romans also identified certain bones as the remains of specific heroes and monsters. Herodotus wrote that a Spartan cavalryman found the bones of Orestes at Tegea. Pausinias reported that the remains of Theseus had been found at Skyros and those of Ajax at Mysia. When the Romans diverted the Orontes River in Syria, they were said to have found the coffin of a giant sixteen feet tall. An oracle told them that it was Orontes, a giant from India for whom the river was named. Pliny described the discovery of the bones of Orion, uncovered by an earthquake in Crete. They were forty-six cubits (sixty-nine feet) long. The stories of various discoveries came to be gathered into a standard canon that each subsequent writer repeated while adding new reports. The most comprehensive, and later the most influential, were those of Pliny and Pausinias.

The early Christians adopted the gloomy tradition of the decay of the world and the historical reality of giants and combined them with the Old Testament narratives: the fall from Edenic perfection and of giants who once oppressed God's chosen people. Several of the early Church fathers wrote about the decline of mankind. The most important of these was

St. Augustine. In *The City of God*, while dismissing the arguments of pagans, he writes:

> They do not believe that the size of men's bodies was larger then than now, though the most esteemed of their own poets, Virgil, asserts the same, when he speaks of that huge stone which had been fixed as a landmark, and which a strong man of those ancient times snatched up as he fought, and ran, and hurled, and cast it—
>
> Scarce twelve strong men of later mould
> That weight could on their necks uphold;
>
> thus declaring his opinion that the earth then produced mightier men. And if in the more recent times, how much more in the ages before the world-renowned deluge? But the large size of the primitive human body is often proved to the incredulous by the exposure of sepulchres, either through the wear of time or the violence of torrents or some accident, and in which bones of incredible size have been found or have rolled out. I myself, along with some others, saw on the shore at Utica a man's molar tooth of such a size, that if it were cut down into teeth such as we have, a hundred, I fancy, could have been made out of it. But that, I believe, belonged to some giant. For though the bodies of ordinary men were then larger than ours, the giants surpassed all in stature. And neither in our own age nor any other have there been altogether wanting instances of gigantic stature, though they may be few. The younger Pliny, a most learned man, maintains that the older the world becomes, the smaller will be the bodies of men. And he mentions that Homer in his poems often lamented the same decline; and this he does not laugh at as a poetical figment, but in his character of a recorder of natural wonders accepts it as historically true.

By the late Middle Ages, the narrative of the aging and loss of vitality of man and the world had acquired a standardized form supported by the familiar list of giants. The giants' roll call still was not closed. Boccaccio reported on a discovery made in Sicily during his lifetime, though he was not present to witness it himself. In 1342, near Trepani, on the western end of the island, a group of workers, digging the foundation for a new house, uncovered a deep cave. They climbed in and found a great grotto where they saw the figure of a seated man of almost unimaginable size. In his hand, he held a staff as large as ship's mast. According to their report, he was two hundred cubits tall (three hundred or four hundred feet, depending on your cubit). The workers hurried back to the village of Erice to share the story of their discovery. Soon, a crowd of three hundred people armed with torches and pitchforks marched to the worksite and entered the cave. Once inside the grotto, they paused, all frightened and awestruck except for one brave man who stepped forward and touched the giant's staff. It disintegrated, leaving only dust and some iron pieces. He then touched the leg of the titan who also turned to dust leaving only three enormous teeth. The teeth were taken to the Church of the Annunciation where they were strung on a wire to be displayed. Boccaccio does not report what happened to the iron. We can safely assume that the local blacksmith took advantage of the free materials. There was some debate over the identity of the giant. Some thought he was Eryx, a legendary early king and founder of the village. Although a demigod himself, Eryx was killed in boxing match with his fellow demigod Hercules who had a bad habit of doing that sort of thing. The opposing and more popular idea was that he was the cyclops Polyphemus and that this was the very cave where he was blinded by Odysseus and his crew. In making that claim, they faced some competition. Over the years, a number of villages in Sicily had discovered a number of caves containing the bones of a number of giants and all had proclaimed their giant to be Polyphemus. Classics scholars, then and now, believed that the *Odyssey* described an itinerary of real places around the central Mediterranean and that Sicily was the home of Polyphemus. Even the average Sicilian peasant knew this and was proud of the history of their island. And if the local giant wasn't Polyphemus, well, enough giants had been found that no one doubted that the island had once been home to a whole race of them.

Almost six hundred years later, the Austrian paleontologist Othenio Abel wondered if there was more to the story than that. In 1862, Hugh

Falconer, one of the first great authorities on the diversity of extinct proboscideans, had presented a paper on the discovery of the remains of a dwarf elephant on the island of Malta. Falconer named it *Elephas melitensis.* In the years after that, other dwarfed species were found on most of the major Mediterranean islands. All of these species, except one, are believed to descended from *Palaeoloxodon antiquus*, the straight-tusked elephant. The exception is a dwarf mammoth that lived on Sardinia. Sicily is especially rich in these fossils, having been home to three different species of dwarfed elephants at different times. Abel thought the skeletons explained the origin of the cyclops myth. Most land mammals share a basic skeletal structure, but proboscideans and humans have some very specific resemblances. The most obvious in the limbs. Both have long straight limbs with short ankles or wrists and five digits. Laying the disarticulated bones of the body of a proboscidean out on the ground, it's easy to form something that looks like an enormous, stocky human. Then comes the problem of the skull. Abel pointed out that the most distinguishing feature of a proboscidean skull—if the tusks are missing—is a huge hole in the middle of the face. This is the nasal cavity, with all of the attachments for the trunk. The true eye sockets are on the sides of the skull and almost unnoticeable. This would make it very easy for an awestruck discoverer to mistake the nasal cavity for the socket of a single huge eye. Other differences in the skulls can be explained by the fact that giants are, by definition, monsters. For example, even if the tusks remain, on *Elephas melitensis* they are very small and can be interpreted as fangs. Finally, add to this the tendency of probosciean skulls to fall apart and the fact that the Greeks didn't encounter elephants until the time of Alexander and you have all of the ingredients necessary to construct a race of one-eyed giants from bones.

When the prominent Basel physician Felix Plater was called to Lucerne in 1584 to care for the ailing colonel Ludwig Pfyffer, he expected to spend his spare time collecting rare plants on the neighboring mountains and visiting with his friend Renward Cysat. He was successful on all counts. The colonel recovered, Plater gathered more than a hundred samples of plants unknown to him, and Cysat had a special treat for him: mysterious bones. His friend explained that seven years earlier, a tremendous storm had buffeted the village of Reyden, a village that Plater had passed through on his way to Lucern. When the brothers of the local monastery came out to inspect the damage, they found that an ancient oak on Kommende Hill had been knocked over. Tangled among its roots were the large bones that Cysat now showed him. Many of the bones were damaged and only a few fragments of the skull remained. Naturally, the workmen were blamed for mishandling them. Plater convinced the city council to let him take them back to Basel for study. From the long bones of the arms and legs and, especially, a digit that appeared to be a thumb, Plater felt confident in telling the Lucerners that they had the remains of a human giant. By his calculations, it stood fourteen strich tall (nineteen feet) in life. Since giants were not part of any local traditions, he believed that it must have lived and died during some prehistoric era before normal humans arrived in the mountains.

Plater hired Hans Boch, an artist who happened to be painting his portrait at the time, to prepare large drawings of the bones and an imaginative drawing of the giant as it must have appeared in life. In Boch's reconstruction, the giant stands with one hand on a dead tree, perhaps an oak like the one he was found under, naked except for a laurel and a girdle of oak leaves. Despite Plater's conclusion that the giant and normal people had never lived together, Boch included a Swiss pikeman, gaping in awe at the giant, for comparison. The Lucerners were delighted, both with Plater's conclusions and with Boch's drawings. The bones were put on display in the city hall and the giant was made the shield-bearer of the city's coat of arms. They had him painted on a tower attached to the city hall with a poem telling the story of his discovery. That wasn't the end of the giant's fame. In the next century, Cysat and members of the city council decided to decorate the three footbridges that connected the two parts of the city across the Reuss River. They hired an artist to paint triangular panels to be hung inside the bridges, attached to the roof trusses. Prominent citizens were encouraged to sponsor panels and in return, their crests would be incorporated into the paintings. Cysat bought panel number one on the Chapel Bridge (Kapellbrücke). For the subject, he chose Boch's giant along with a poem that he composed. Later, the Jesuit Athanasius Kircher adapted Boch's drawing to illustrate the relative sizes of famous giants. In his well-known version, six smaller giants stand in the shadow of Boccaccio's titan. All seven have the posture and attire of Boch's giant. The Reyden giant stands next to Goliath as the second smallest. Habicot was twenty-seven when the tree in Reyden was blown over and was quite familiar with the conclusions of the Swiss doctor.

The Renaissance and Reformation brought with them strong intellectual challenges to the historical reality of giants and equally strong defenses of it. Plater and Cysat's embrace of the giant of Reyden amounted to taking sides in an emerging debate. It was impossible for either Protestants or Catholics to come right out and say that the Bible was wrong about giants, even if they secretly believed it. The challengers crouched their arguments in terms of certain passages in the Bible needing to be read as allegory. The giants of old, they wrote, were giants in deed (good or evil) rather than giants in stature. When Theodore Zwinger wrote in 1565 that "although Scripture [uses the word giants], theologians nevertheless prefer to interpret these passages allegorically," he was ahead of his time by a good century. No such consensus existed. One year earlier, a work was published by the poet and military engineer Girolamo Maggi (using the Latinized name Hieronymus Magius) in which the author struggled with ways to reconcile the evidence and non-evidence of giants and the aging of the earth. On the one hand were the authoritative writings of the Bible and respected authors of antiquity. On the other hand were the results of his personal research. He had examined Roman and Etruscan tombs and weapons from the Punic Wars, almost two thousand years earlier, and found them to be no bigger than those of his day. Maggi began by reasoning that many ancient accounts were exaggerated; certainly, there had been giants, just not really gigantic giants. Next, he took a rather creative approach to the doctrine of the decline of the world by suggesting that the loss of vitality didn't happen at a consistent rate in all places and times. Thus, there could still be large numbers of extraordinarily large people in China and Patagonia while the Italians had remained the same size for the last two millennia.

The first unqualified denial of giants came at the end of the decade from the pen of the Flemish physician Jan van Gorp (Joannes Goropius

Becanus) in his book *Origins Antwerpianae*. According to a popular legend, the future site of his city had once held a bridge across the River Scheldt guarded by a giant named Druon Antigoon. If anyone refused to pay their toll he would chop off one of their hands as payment and throw it in the river. He continued this unpopular practice until the arrival of a Roman soldier named Brabo who chopped off one of Antigoon's hands and tossed it into the river while the giant bled to death. Brabo then became the namesake of the province of Brabant and the city that grew up next to the bridge took the name Antwerp from the Flemish words "hand werpen" which roughly translates as "hand tossing." The part of the story that Gorp took issue with was the meaning of the word "giant." Much of his work involved the construction of elaborate and fanciful etymologies. During his researches, he had been amazed to discover that the dialect of Flemish spoken around his home in Antwerp was the exact same language that had been spoken by Adam and Eve in Eden. This made it the oldest and most noble language on earth. It escaped the confusion of the languages at Babel because the tribe of Gomer, the ancestors of the Flemish, had already trekked off to Northern Europe by the time the tower was allegedly built four thousand years before his time. It had remained pure and unchanged during all those centuries. Through linguistic contortions, he proved that the Latin word *gigant*, was a contraction of the Flemish *wijt gehant* which he gave the meaning of *qui manus habet longe lateque extensas* [He has his hands stretched far and wide]. The name of the antediluvian giants in Genesis, Nephilim, comes from the Hebrew word *naphal* which he said was a cognate of the Flemish word *val* meaning "downfall." From this, he concluded that the word "giant" had originally been used to describe a people who were in rebellion against God and who were punished for it. Large bones that had been found in the vicinity of Antwerp, he dismissed as nothing more than the bones of elephants brought north by the Romans or perhaps sea monsters that had gotten lost and swum up the river.

Though Gorp was not the first to propose an allegorical interpretation of ancient giants, the blunt and uncompromising nature of his attack demanded a response. Several were written at the time and continued to be argued over the next half century. Even Sir Walter Raleigh, who spent his time while in the Tower of London writing a history of the world, felt compelled to write a defense of giants. Jean Chassanion wrote an entire book to refute Gorp. Chassanion tried to overwhelm Gorp with an appeal to authority and the sheer volume of his evidence. His book was primarily the standard list of giants from the Bible and esteemed ancient writers and well known recent discoveries, such as the giant of Erice, supplemented by a few previously unmentioned giants that he uncovered in his research, including a giant tooth that he had seen with his own eyes. The book went through three editions. Chassanion did make one departure from his predecessors; he did not keep the theme of the decay of the world. His giants were departures from normal humanity, which had always been the size of modern men.

This was the state of European thought on giants when Mazurier picked up his bones and departed Paris, leaving Habicot the job of defending Theutobochus.

The response to Habicot came in December, just before the end of the year, in the form of an anonymous pamphlet entitled *Gigantomachie pour respondre à la Gigantostologie* (Gigantomachie, a response to Gygantosteologie). The author was given as "a Student in Medicine" but everyone knew it was Jean Riolan, an influential member of the guild of physicians and son of a famous anatomist of the same name. Riolan never indicates that he saw the bones. His arguments are entirely directed at Habicot's pamphlet. On the question of giants in general, Riolan matches citations of respected authorities of antiquity who argued for the reality of giants with his own list of respected authorities of antiquity who argued against the reality of giants. As to the identification of the bones as those of Theutobochus, he states that this is either a fraud or a credulous mistake on Habicot's part. Riolan is at his strongest when he delves into the purely anatomical details of Habicot's identification. After a lengthy and pedantic section focusing on the number of bones in the human body, he gets to the meat of the matter; the measurements that Habicot provides do not match the proportions of a normal human body. The skull is too big and the chest is too deep to fit with the named height of the skeleton. He concludes that the bones must have come from an elephant. Although he, like Habicot, admits he has never seen one, he thinks the bones are roughly the right size according to some descriptions he has read. Riolan does not limit himself to responding to Habicot's arguments or to advancing new ones related to questions surrounding the bones. He attacks Habicot's competence as an anatomist, the value of his previous works, his style as a writer, and even his spelling. He frequently throws out the words "ignorant," "ridiculous," and "inept." Not satisfied to go after Habicot himself, Riolan expands his attacks to the entire guild of barber-surgeons and to its most famous practitioner, Ambroise Paré. Paré was considered one of the titans of French surgery and had been Habicot's teacher, making this an especially bitter line of attack. Even his pseudonym "a Student in Medicine" is an insult; it carries the message that even a novice should see how wrong Habicot is.

Why was Riolan such a jerk? The medical professions in seventeenth-century France were bitterly divided. Physicians, barber-surgeons, and apothecaries each had their own guilds and jealously guarded their prerogatives. Outside the guilds, tooth pullers, bone setters, midwives, and oculists also practiced professionally, though with a lower social status. Paris was the center of education in France and both the physicians and barber-surgeons had schools there. The University of Montpellier, where the doctors had authenticated the bones, had its own school of medicine that competed with Riolan's University of Paris. The social ranking of physicians and surgeons was the opposite of the way most Americans use the words today. Physicians had a far superior social ranking than surgeons. Physicians not only practiced medicine, they taught it, they studied the great writers of the past, and they communicated in the purest Latin. By contrast, surgeons were manual laborers of the human body who spoke vulgar French. Physicians treated diseases; surgeons repaired injuries. Only physicians could order internal medicines to be prepared by apothecaries; surgeons were limited to preparing ointments and topical treatments. The boundaries were a constant source of struggle with new charters being issued several times a century. The barbers were at the bottom of the totem pole in this struggle. Their medical role was to provide first aid in the villages. Sometimes they were merged with the surgeons and sometimes they were separate. Habicot and Riolan were both active participants in this political-professional struggle between the guilds. At the time, Habicot was working to once again separate the barber-surgeons from a more professional group of pure surgeons. Riolan acted from time to time as the inquisitor-general for the Faculty of Medicine, accusing and prosecuting physicians who dared use unapproved treatments. Habicot was a prolific writer on medical issues who published a practical manual on medicine. In describing his own discoveries and innovations, he appeared to cross the line into teaching and this was something Riolan could not tolerate.

The battle over Theutobochus continued off and on for five years. Four months after his original response, Riolan published a second anonymous attack, titled *L'imposture descouverte des os humains supposés, et faussement attribués au Roy Theutobochus* (The fraudulent discovery of alleged human bones falsely attributed to King Theutobochus). Even though the whole medical community knew he was the author, he still published anonymously, even going so far as to compliment the author of *Gigantomachie.* In this book, he expands on his anatomical critique of Habicot's description and expands on his historical arguments. He does not go so far as to accuse Habicot of fraud but does call him naïve for being fooled by the real charlatan, Mazurier. A full quarter of his pamphlet is a direct attack on Habicot's competence as a medical practitioner through a scornful rebuttal of an earlier treatise the surgeon had written on the muscles of the diaphragm. A significant difference between this pamphlet and *Gigantomachie* is that Riolan no longer claims the bones belonged to an elephant. Having had a while think about it, he doesn't think these could have survived for seventeen centuries, as Tissot claims. Animal bones rot in the ground; only saints' bones, being pure and incorruptible, last for centuries. The Theutobochus bones must be figured stones or "sports" or "jokes" of nature, natural mineral productions that only happen to look like bones. The word Riolan used to describe such a thing was *fossile.*

The word *fossil* has undergone a lot of evolution since Roman times. The original Latin root, *fodere,* is a verb meaning "to dig." From there the word evolved to become a noun, *fossa,* describing the excavation itself—a ditch, trench, or moat—and continued to be used in that sense by military engineers into modern times. By the Renaissance, the form *fossile* had appeared as an adjective or noun meaning "a thing dug up." At first, it meant anything of interest or value found in the earth. Metal ore, gemstones, crystals, opals, and other stones containing images, and bones or stones that looked like bones were all dubbed fossils at this time along with human artifacts like pottery and hidden treasures. None of these were quite as passive as we think of them today. Figured stones could grow and move directed by various forces, such as the astrological influences of the stars. Many examples of such mineral growth could be observed. Crystals could be grown from a solution of salt. Sometimes stalactites and stalagmites could be measured and seen to become longer over a period of years. Mineral springs were known to encrust objects with calcium making them look like stones. Sand and small pebbles grew in some peoples' kidneys. And every gardener, even today, knows that small rocks "grow" in vegetable patches over the winter. This generative power of the earth, sometimes called *vis plastica*, made it reasonable for people to think that not only could one bone-shaped stone form in isolation, but that several could form together giving the false appearance that they had once belonged to a living person or animal.

If the tusks of the giant found on Marquis Langon's land had survived, it would not have challenged Riolan's conclusion in any way. Since antiquity, it was known that ivory could be found in the earth. Theophrastus, Plato's student, wrote that a type of ivory "with white and dark markings" could be dug up. Theophrastus's, work, *On Stones*, survived and influenced writers well into the Renaissance. In Riolan's day, this substance was known as *ebur fossile* (fossil ivory). Pliny, and other Roman writers were happy to accept fossil ivory as the product of elephants. Elephants, being the wisest of animals, had many human characteristics, including burying their dead. No one believed that wild elephants had lived in France, but Hannibal and other armies had passed through. During the Middle Ages, this Classical tradition of thought about buried ivory began to be replaced by another set of beliefs. During the Renaissance, many, especially of the upper classes, believed fossil ivory to be the remains of unicorns, and one of the most precious of medicines. Riolan mentions the fact that buried ivory was often sold as unicorn.

Unicorn lore began not as a legend, but as a rumor. Ctesias of Cnidus left the Greek world in 416 BCE to become the personal physician to the

Persian emperors Darius II and Artaxerxes II. He stayed abroad for seventeen years. When he returned to Greece, he wrote books based on what he had learned there. Though he never left the royal court, which stayed in the western part of the empire, he spent many evenings with merchants and travelers listening to their stories of lands to the far east. Ctesias did not have a great reputation among the literati of subsequent generations. At best, he was called gullible and at worst a liar. Only fragments of his books have survived leaving us little from which we can form our own judgment about his credibility. Not everything he wrote was dismissed out of hand. Some of his natural history was cautiously accepted and commented on by serious writers like Aristotle and Pliny. One of the animals he wrote about was the *monoceros* or unicorn. He described the unicorn as a type of wild ass native to India, as large as a horse and with a single horn growing from its forehead that was white near the animal's face, black in the middle, and bright red at the tip. It was swift and powerful and almost impossible to catch. So far, there is nothing particularly fantastic about this animal except for the horn. The horn was special: "Those who drink out of these horns, made into drinking vessels, are not subject to convulsions or to the holy disease [epilepsy]. Indeed, they are immune even to poisons if, either before or after swallowing such, they drink wine, water, or anything else from these beakers."

A century later, a second Greek writer described another type of unicorn. Megasthenes traveled to India on a diplomatic mission for the Selucid dynasty then ruling Persia. Despite being in India, he also never saw the animal and had to rely on descriptions provided by others. His animal was called a cartazon and lived in remote mountainous regions. It was considerably more exotic than Ctesias's animal: "This creature is as large as a full-grown horse and has a mane and soft yellow hair. It is provided with excellent legs and is very swift, for its legs resemble those of an elephant being without joints. The tail is like a pig's." The idea that elephants do not have knees was denounced by Aristotle, but continued to repeated into the Renaissance. Megasthenes added a detail that would become important fourteen centuries later. The horn, he said, grows "not symmetrically but with natural twists." He does not mention any medicinal uses for the horn or any other part of the cartazon.

Modern writers tend to dismiss both descriptions as being of rhinoceroses, some kind of gazelle that is only ever seen in profile and from a distance, or a garbled mix of several animals. Other than being the size of a horse and having the single horn, the two descriptions don't resemble each other. The body of Ctesias's unicorn is nothing more than a distinctly colored wild ass while Megasthenes's is a composite of other animal parts like a gryphon—a lion with the head, wings, and talons of an eagle—or a manticore—a scorpion-tailed sphynx—(which he also describes). It's always possible that Ctesias's and Megasthenes's informants were just pulling their legs. Attempts to match the descriptions with animals in classical Indian literature have so far been fruitless, making it unlikely that we will ever know what they were told. The unicorn did not exist in the popular mind in the ancient world. It does not appear in poetry, drama, or art. For the six centuries after Ctesias, it appears only a handful of times and those were all in the context of learned men writing about natural history. They wanted to know exactly what kind of an animal the unicorn was. How should it be grouped with other animals? Was it the only animal with one horn? Pliny listed six single-horned animals including the rhinoceros and a type of Arabian gazelle. Julius Caesar calmly reported a single-horned animal in the Hercynian Forest, which stretched eastward from Germany and was the edge of the known world for Romans.

This changed with the spread of Christianity. Sometime in the early years of the Church—possibly as early as the second century—a type of book called the *Physiologus* began to circulate. These books were collections of short essays on various animals, and sometimes plants, made up of quotes from the works of respected ancient authors followed by moral lessons exemplified by each animal. In the late Middle Ages simplified versions of the *Physiologus* would be called bestiaries. It was through these books that the unicorn was introduced to the masses. The earliest of these described an animal much smaller than those of Ctesias's and Megasthenes's. It looked like a kid goat but was ferocious and dangerous. It could only be captured by a virgin who would sit in a meadow. The unicorn would come to her and lay its head in her lap. After that she, sometimes assisted by a huntsman, would take it to the king. This drama was immortalized in friezes and tapestries. In time the unicorn grew in size eventually becoming once again the size of a horse and it was usually portrayed as the horse-like creature that we know today. The horn, called an alicorn, grew even faster. Ctesias (fourth century BCE) wrote that the horn was one cubit (eighteen inches) in length. Pliny (first century CE) wrote that it was three feet long, Isidore of Seville (seventh century) wrote that it was four feet long, and Albertus Magnus (thirteenth century) made it a whopping ten feet long. Skeptics pointed out that a unicorn would have to be as big as a ship to support such a horn. Albertus wasn't speculating when he specified that size. By his day, Norse colonists had begun bringing narwhal horns to Europe from Greenland and other cold seas. Norway was very secretive about its fishing grounds and knowledge of the animal that produced the horns was slow to spread. These horns had spiral grooves exactly like the ones artists had been portraying for a thousand years and easily could be ten feet long.

As improbable as ten-foot horns sounded, there was a growing market for alicorns making people want there to be bigger horns, which could be sold for more and which were impressive trophies for their owners. In the late Middle Ages, Europe began to descend into a poison panic. Ctesias's statement that alicorns were both protection against and a cure for poison led the rich and powerful to seek them out to an unprecedented degree. Poison, as a tool to get rid of political opponents, relatives, and generally inconvenient people is as old as politics and wills. Essays on poison and antidotes to poison have been found among the earliest documents of all of the civilizations of the Old World. Entire industries have grown up around manufacturing poisons, poisoning, preventing poisoning, and curing the poisoned. Throughout history, some people have had very good reasons to fear being poisoned. However, there have been times when poison fears have gripped societies, fears that far overestimated the abilities of poisoners and the desirability of many who imagined themselves to be targets. Given the state of the medical practice, disease theory, and forensics in Renaissance Europe, it was easy for people see poisoning in every unexplained or sudden death. As in any irrational panic, mercenary minded people were ready to exploit public fears to their advantage. Some used mysterious deaths as an excuse to incite mob action against their personal enemies or against outsider groups. Occasionally this led to pogroms against Jews and Roma. The less murderously inclined saw in these fears a way to turn a fast buck. Some sold manuals and tools for poisoners. Others sold antidotes and protective amulets, such as unicorn horn.

By the peak of the poison panic in the mid-sixteenth-century, tiny fragments labeled as alicorn were selling for ten times the price of gold. A large piece of horn could command twice that or even more. A complete and well-shaped narwhal tusk was something that only kings and cardinals could afford. To meet the demand, any intriguing-looking bits of bone were sold as alicorn. Paranoid buyers snapped up white stalactites because their tapering shape could be mistaken for a horn. Foreign craftsmen were reputed to know the secret of straightening elephant or walrus ivory. It's almost certain that some bits of mammoth ivory were drafted into service as alicorns. Finally, fossils, plain white rocks, and even vials of water that once ostensibly been touched by an alicorn could be sold for ridiculous prices. At the same time that the demand and price of alicorn were inflating, so were its reputed medical properties. By the time the panic peaked, alicorn provided "effectuall cure these diseases: Scurvy, Old Ulcers, Dropsie, Running Gout, consumptions, Distillations, Coughs, Palpitation of the Heart, Fainting Fits, Convulsions, Kings Evil, Rickets in Children, Melancholly or Sadness, The Green Sickness, Obstructions, and all Distempers proceeding from a Cold Cause." Reports even surfaced that the horn could raise the dead. Well-meaning rulers ordered tests to protect buyers from fake alicorn (a Renaissance version of the FDA). These usually involved waving the purported alicorn at something like a poisonous snake to see if it was repelled or poisoning pigeons and seeing if it could revive them. A surprising number of products passed the tests.

The alicorn's very commercial success doomed the unicorn. Too many people were making a close study of the unicorn and nothing about it could stand up to extended scrutiny. Physicians questioned the idea of a universal antidote. Nature, they believed, was composed of paired opposites, hot versus cold, wet versus dry. How could the same medicine counteract a wet poison and also a dry poison, a warm poison, and a cold poison? Naturalists hunted the world for the unicorn animal and found hints and claims, but no actual unicorns. A mere two years after Girolamo Maggi challenged the historical realty of giants, similarly Andrea Marini challenged the reality of unicorns. In his *Discorso contra la falsa opinione dell' Alicorno* (Speech against the false belief in Unicorns), published in 1566, Marini asserts that the traditions about the unicorn and the medicinal properties of the alicorn are so wide and varied that they could not possibly describe one thing. He suspects that the horns in the north, especially those found by the English, must come from a marine animal. As for its medicinal properties, Marini stated that though alicorn might be effective against fevers and poisonings coming from a cold, dry cause, it is no more effective than a common stag's horn. More than anything, he feels that purchasers of alicorn are being taken advantage of by unscrupulous merchants. Marini was almost immediately answered by Andrea Bacci, a rising star and future physician to the pope. Bacci argues that the traditions surrounding the unicorn are confused because it is so rare and not native to Europe. He makes the traditional argument from authority with an added bit of class snobbery: if the unicorn was merely a superstition of vulgar, common people, we might have cause to question it, but as the greatest minds of the past believed in it, we do not. As to its medicinal properties, this is proved by its very rareness. God made it rare because it is special; things special to God must have extraordinary properties. For the time being, Marini and Bacci's debate didn't change popular opinions about the unicorn, but it did set the terms of the debate as it would develop in the next century.

One other important writer commented on the unicorn before the century was over. This was none other than Habicot's teacher, Ambroise Paré. Paré was the surgeon to the court of France. In that position, he had to be careful expressing his doubt. His employers had expended exorbitant sums on alicorns and he had to be careful not to say anything that would embarrass them. He could have stayed quiet, but he was of the same mind as Marini, in that he hated charlatans and itched to expose frauds. In 1580, Paré was called upon to tend to Chevalier Christofle des Ursins who had an infected injury caused by a fall from his horse. Des Ursins was interested in the specifics of his care and asked why Paré didn't use famous medicines such as dried mummy or ground alicorn. Paré was able to reject the first treatment by saying it was improper for a good Christian to consume the flesh of a dead pagan. For the latter, he explained that his own experiments, which had involved poisoning a condemned prisoner and giving him a healthy dose of alicorn, had shown that its properties had been highly exaggerated. Des Ursins was so impressed by Paré's knowledge that he insisted he write it out for the good of humanity. Paré repeats many of the same arguments that Marini used and supplements them with additional experiments that don't involve killing servants. In the end, he concludes that while there might be a rare, one-horned animal found in nature, the alicorn had very little use in medicine, if any. He lamented that his main purpose in using it was because his customers demanded it. If he failed to prescribe it and the patient died, he would be in danger of lawsuits or worse.

While the debate over alicorns and unicorns continued to simmer, Habicot for his part refused to respond to Riolan about giants and a full year passed before the debate flared up again. In March 1615, an anonymous pamphlet appeared titled *Discours apologétique touchant la vérité des Géants* (Apologetic speech concerning the truth about Giants), generally credited to Charles Guillemeau, the personal surgeon to the king. As he had been attending the king since 1612, he probably had seen the bones when they were brought to the palace. Being a surgeon, the author opens with a full-throated defense of the guild. On matters more closely related to the bones of Theutobochus, the author brings down a pox on both houses in the debate, but a much bigger pox on Riolan's house. His argument is that the truth of giants is unassailable, being based on Holy Scripture and the wisdom of the church fathers. Riolan, he writes, committed sacrilege by denying the reality of giants. Habicot's sin, he says, was getting involved with the charlatan Mazurier. Even this mild criticism from his own side was too much for Habicot. The following month he issued a thirty-six-page response in which he repeated his previous arguments and finished with with his own defense of the guild. The most significant aspect of this response was his confidence that his original conclusions were correct. While Riolan had been confident that the bones were not those of a giant, he had wavered over just what they were. Habicot was unchanging in his faith that the bones were those of a human giant and that the giant had been Theutobochus. After that, the duelists were quiet for two years until, in 1617, something provoked Riolan to return to the field. In fairly short order he published two brief pamphlets that were nothing more than collections of cheap shots at Habicot. Habicot fired back and Riolan issued one more collection of insults, but the end was near. In 1618, they each published small books summing up everything that had been said before and laying out their final arguments in detail.

Riolan, finally leaving aside his anonymity and publishing under his name, was first with *Gigantologie: Histoire de la Grandeure des Geants* (Gigantologie: The history of the Greatness of Giants), probably published near the beginning of the year. He attacks the idea of the aging of the world but is careful not to suggest that the idea came from the Bible. He credits it to the Greek philosophical school of the Epicureans. This identification is safe because the Classical Greeks were, after all, pagans. He tells his readers that the idea of the aging of the world does not make any sense. Old age is marked by sterility, not by shrinking. He allows that people before the Flood might have been a little taller than modern people—say, nine or ten feet tall. Genesis says there were giants then, but since the Flood we've all been about the same size. He also allows that from time to time there have been people of extraordinary height, like Goliath and Charlemagne, but that they have been rare and that their height was exaggerated by the writers of their time. The maximum height possible for a human is fifteen feet, eighteen tops. Next, he analyzes the possible explanations for reports of giants. Bones and teeth found near the sea could come from whales, nereids, sirens, or unknown monsters. There have been many reputable accounts of such discoveries. He counts St. Augustine's tooth among these. They could come from elephants brought by Hannibal or the Romans. Elephants are reported to have knees similar to a human's as well as five toes. Lacking a skull, the bones could easily be mistaken for a human giant by a person without his advanced anatomical knowledge. They could be "fossils," in the same vein as stalagmites or crystals. He describes various reported oddities taken from the earth shaped like a brain, praying hands, a woman's "shameful parts," and bones. He mentions an episode of German workers digging a foundation who came across objects that looked exactly like pots made by a master craftsman. That's an archaeological site that we today will sadly never get to examine. He refuses to choose among the possibilities. His message is that the bones could be anything at all—except a giant. To drive his points home, he concludes with an abridged version of his 1614 pamphlet *L'imposture descouverte.*

Later in the year, Habicot presented his rebuttal, titled *Antigigantologie: Contre Discours de la Grandeur des Geans* (Antigigantologie: Against the Discourse on the Greatness of Giants). His arguments are the exact inverse of Riolan's. Riolan examined the possible explanations for giant bones in the earth and concluded that any of them was possible except that they were a human giant. Habicot examined the possible explanations for giant bones in the earth and concluded that none of them was possible except that they were a human giant. Monsters are defective and against nature. They have two heads or six feet. The bones of his giant were perfect. They had the right shape and there were the right number of them. This claim was helped by his never having seen the skull, only teeth. The bones could not come from a whale. Whales are much bigger than his bones and have no teeth or feet. The bones could not have come from an elephant. The bones are bigger than an elephant's. Elephants do not have heels and their teeth are different from his giant's. As for fossils, he argues semantically. Bones, in all of their complexity, do not fit into any of the categories of fossils described by the great authorities. They are not crystals. They are not metals generated by the influence of the planets. They are not shapeless stones or rocks containing superficial resemblances to the outer world. He adds one more argument not based on anything Riolan has said but based on old dualistic properties: of all the dualistic properties an object can demonstrate—hot/cold, wet/dry, heavy/light—only bones have the combination that makes them recognizable as bones. Having laid out these arguments in the first part of his book, the second part is a point-by-point annotation and response to Riolan's *L'imposture descouverte.*

Habicot had the last word. Between his own arguments and his line-by-line response to *L'imposture descouverte,* he inserted the texts of three letters. The first is from Mazurier in 1614, saying that he had forwarded to Langon Habicot's request for some documentation of the discovery, particularly the notarized statement to Constable Assalin. The second is an undated note from Langon to Habicot in which he assures the surgeon that the doctors of Grenoble and Montpellier had authenticated the bones as coming from a human giant. He promises to send the statement, a coin, and everything else the king had requested just as soon as he returns from his current business trip. The third letter is from Mazurier to Habicot dated June 1618. In this he reports that he heard Langon was finally on his way to Paris with bones and the notarized statement to show the king. He does not say that Langon was bringing any parts of the tomb. Riolan never responds. Did Langon provide the king with everything Mazurier said? If he did, no records have survived. From Riolan's perspective, it wouldn't have mattered. The letter from Langon confirming the reality of the discovery and that Marshal Lesdiguieres was satisfied that the bones were those of a human giant was enough to end the dispute. Riolan could claim that, as a doctor, his opinions were more valid than those of a mere surgeon, but he could not challenge the word of Langon and Marshal Lesdiguieres. The marshal was one of the most powerful men in the kingdom, and Langon was no common landowner, he was the Marquis de Langon, Baron d'Uriage, and Lord of Saint-Julien, Montrigaud, and other places. In the pecking order of the kingdom, Riolan was below both of them. It would have been dangerous to pick a fight with them.

In the course of these raging debates, one thing both Habicot and Riolan had admitted was that neither had ever seen an elephant. They could not perform comparative anatomy because they had nothing to compare the bones to. Since Roman times only a handful of elephants had been seen north of the Alps. The illustrations in bestiaries were fanciful and wildly inaccurate, based on the writers of antiquity and not on the observations of recent travelers. The entry on elephants in Konrad Gessner's influential natural history, *Historiae animalium*, showed a big gray animal with an approximately correct body shape, giant ears like folding fans, tusks that pointed straight down like fangs, and a trunk like a vacuum cleaner hose. The closest thing they had to an anatomical essay was a letter written by Pierre Gilles, a member of a French embassy to the Ottoman Empire, to the bishop of Syria in 1548 describing the autopsy of a young elephant in very general terms. There were no drawings of an elephant's skeleton available to Habicot or Riolan.

This situation had already begun to change during their lifetimes. Stories of monarchs exchanging animals as gifts date back almost to the beginnings of recorded history. Most educated Europeans of the day would have been familiar with the elephant Abul-Abbas gave to Charlemagne by the Abbasid caliph Harun al-Rashid and the elephant given by Louis IX of France to Henry III of England in 1255, which became the namesake of the Elephant and Castle district of South London. Animals that are strange and wonderful, powerful and regal reflect on the greatness of the givers, the vastness of their realms, and their ability to control such things. It's no coincidence that one of the first acts after Nixon and Mao began the process of regularizing diplomatic relations between their countries was an exchange of exotic animals, pandas from China and musk oxen from the United States. As the Portuguese began working their way down the African coast in the fifteenth century, looking for the sea route to India that they would find at the end of the century, they encountered and brought back exotic animals as tribute to the monarchs and patrons who sponsored them. The earliest animals were kept in the menagerie of the king. In 1514, Manuel I of Portugal sent a white elephant named Hanno to Pope Leo X as a coronation gift. Hanno died soon after. The next year, Sultan Muzafar II of Cambay sent Manuel a rhinoceros, the first seen in Europe in more than a thousand years. Manuel promptly regifted the rhino to the pope as a replacement for Hanno. Sadly, it drowned in a shipwreck before reaching Italy. The Portuguese monarchs handed out other elephants during the sixteenth century. Because they had only broken bones to work with, seeing a live elephant would have been of limited use to Habicot and Riolan, though it would have served to establish its general size and proportions. What they really needed was a detailed anatomical study with illustrations. This wouldn't appear until the last quarter of the century, long after both men had died.

In the winter of 1613-1614, Nicolas Claude Fabri de Peiresc passed through Paris and took note of the Theutobochus controversy. He could hardly have avoided it. Peiresc was one of the central figures in the Republic of Letters. It seems as if he corresponded with everyone about everything. Seven thousand of his letters have been published and boxes of them remain uncatalogued in various collections around Europe. He was also an accomplished scientist (he discovered the Orion nebula) and keen observer of the world around him. Peiresc arrived in Paris after Mazurier departed and never personally saw the bones. He only knew the controversy through the war of pamphlets. Among his surviving studied letters, the earliest indication of his interest in the matter was seventeen years later. In 1630, Thomas d'Arcos wrote to a common friend describing the discovery of the bones of a giant near Utica in modern Tunisia. This was the very place where St. Augustine's giant tooth had been discovered. Once again, all of the bones turned to dust when the workmen tried to remove them from the ground. Only two teeth survived. Early the next year, d'Arcos acquired one of them and sent it to Peiresc. Peiresc thought the tooth looked familiar, but couldn't remember where he had seen one before. At first, he thought it might have come from a marine monster like a whale or a hippo.

But then, it all came together for Peirsec. In 1626, Claude de Lorraine, duke of Chevreuse, had acquired an elephant, probably from the Dutch who, at the time, were wresting control of trade in the Indian Ocean from the Portuguese. When the elephant arrived, Pieresc heard about it and wrote a friend in Paris asking to compare some of Gilles's descriptions to the live elephant. Pieresc wanted to know more. Two years later Claude rented the elephant to George Pierre, an officer who had served under him in one of the many wars of religion in that century. The contract they drew up gave Pierre the right to travel throughout the kingdom showing the elephant for two years beginning in May 1630. In November 1631, Peiresc heard that the elephant would be passing near Belgentier, his estate, and seized the opportunity to learn more. He convinced Pierre to drop by and stay for a few days. He spent the first part of the visit feeding the elephant treats so that it would become comfortable with him, and because he was curious about its diet. Later, Pierre demonstrated that it was gentle enough to let him examine its teeth. In a letter to Pierre Dupuy, the royal librarian, he described the scene: "I was curious enough, or (rather) mad enough, to introduce my hand in its mouth, and to catch and to feel one of its molar teeth, to better recognize the shape." The elephant even let him make a wax cast of one of its upper teeth. He immediately recognized the shape as being the same as the one d'Arcos had sent him earlier that year. When writing to Dupuy he included a sketch of d'Arcos's tooth.

Peiresc now became curious about other reports of giants' bones. Could they all have come from elephants? He wrote to his correspondents in various parts of Europe where giants had been reported and asked for descriptions of the bones and the circumstances of their discovery. In August 1634, he received a package from Dr. Nivolet of Saint-Marcellin near the Langon estates in Dauphine. Nivolet recounted the Theutobochus story from memory. Despite being a physician, he was convinced that Habicot had been correct and that the bones belonged to a human giant. Nivolet was able to do more than retell a story; he visited the Chateau Langon and interviewed the marquis's widow. The widow Langon showed Nivolet the remaining bones and medals. When Nivolet left, she gave him some bone fragments and one of the medals, all of which he forwarded to Peiresc. None of the bone fragments was large enough to identify other than to determine that they were unquestionably organic in nature and not "fossils." Peiresc was an avid numismatist with a collection of coins and medals numbering 18,000 at the time of his death. After a short search, he was able to determine that the MA on the medals stood for the city of Marseilles, not for General Marius. Based on his historical knowledge, he was able to demolish the rest of the Theutobochus story. In particular, he was disturbed by the description of the tomb. Who would have built such a tomb? Theutobochus's people were either killed or enslaved by Romans. Even if a small number had escaped those fates and paused to entomb their leader, they would not have inscribed his name in Latin. The Romans, he acknowledged, were a chivalrous people and might conceivably have built a tomb for a noble foe, but they would not have used vulgar brick or built it on a low sandy spot.

The widow Langon would not part with the teeth, but based on Nivolet's description of the weight and size of the teeth alone, Peiresc was confident that the bones were those of an elephant. A second "giant" found in the vicinity of Chateau Langon that same year further convinced him that these were elephants and not historical humans, as only one Theutobochus was known to exist, not a nation of them. Peiresc remained interested in giants for the rest of his life. Even as he lay on he lay on his deathbed, he dictated letters to friends in Italy begging them to go to Sicily and investigate Boccaccio's giant. He would not have been as confident in his conclusions about the Theutobochus bones if he had seen a tooth or if Nivolet had given a detailed description of one. The Theutobochus teeth were enamel-covered, four-cusped molars, much like a human's, as opposed to the loaf-shaped teeth, with a wash-board surface that he had seen on the teeth of the living elephant and one sent to him by d'Arcos. The fame of the bones led to them being saved by the Langon family for centuries. In 1984, Léonard Ginsburg tracked down the last known Theutobochus bone, a tooth, and identified it as the third right premolar of a *Deinotherium giganteum*, a strange-looking proboscidean from the mastodon side of the family. He also examined the drawing of d'Arcos's African tooth and identified it as the upper left molar of one of the ancestors of modern African elephants.

In 1645, the twenty-seventh year of the Thirty Years War, Swedish armies inflicted a devastating blow on the imperial forces in Bohemia and swept into Austria with the aim of capturing Vienna. The imperial capital was not prepared to give up easily. Both sides soon found themselves digging in for a long siege, negotiating with allies for support, and building fortifications and counter-fortifications across the countryside. Upriver from Vienna, in the Krems district, while digging trenches, a group of Swedish soldiers discovered the bones of a giant. The discovery took place on St. Martin's Day, November 11. The soldiers had been ordered to build a series of defensive fortifications around an old tower at a place called Laimstetten. The winter was not making their job any easier. Rain and groundwater filed the trenches. To deal with this, the engineers in charge ordered the men to dig a series of deep drainage ditches down the hillside. It was in one of those ditches, at a depth of three or four klatters (eighteen to twenty-four feet), in a layer of yellowish soil that smelled of decay, that they ran into a cache of enormous bones. The most impressive of the bones are described as being a skull as large as a medium-sized table, arms as thick as an average man, a shoulder-blade with a socket large enough to hold a 24-pound cannonball, and teeth weighing up to five pounds. Someone in charge ordered the diggers to save the bones so that they could be sent to learned men in Sweden and Poland for study. Once again, many of the bones, including the skull, fell to pieces as they were brought out. Once again, the workers were blamed for mishandling the bones. Two more giants were uncovered in the trench but, with a war to be fought, they were left in the trench and nothing more was said about them. After the Swedes left Krems, fathers from the local Jesuit monastery combed over the site and recovered a few good teeth that the Swedes had missed along with a cartload of unidentified fragments. They sent the best tooth to Emperor Ferdinand III, "an artful and intelligent man," in Vienna.

The story of the Krems bones were first told just two years later in *Theatrum Europaeum*, a journal of contemporary German events. *Theatrum* was published, edited, and illustrated by the Merian family in Frankfort. The account is short, only 350 words, and written by Johann Peter Lotichius, the personal historian to Emperor Ferdinand. Lotichius tells us the facts of the discovery and the disposition of the bones but hazards no guesses as to what they might have been. Almost thirty years later, Peter Lambeck wrote a catalog of the imperial collections. He uses the Krems tooth as the opening to engage in an extended digression on giants. After repeating the most recent opinions on the truth of giants he comes to the conclusion that we must still believe in giants because the Bible tells us they existed, but that this tooth in particular did not come from one. He also rejects the theory that it came from a Carpathian dragon. He's inclined to believe it came from an elephant. He then moves on to a two-headed chicken that apparently shared a shelf with the tooth.

Lotichius's short account of the discovery is the only one we have. All later accounts are based on his. But it's not the only evidence we have. Matthew Merian, the then-current head of the family publishing business, was an excellent engraver, as was his father before him and his son after him. Merian thought the story interesting enough that he visited the monastery and made a detailed illustration of a tooth the Jesuit brothers kept. It is the only illustration in that volume of *Theatrum* that is not a portrait or map. The illustration is of such fine quality that, centuries later, it is possible to identify it as coming from a young mammoth.

In 1911, two hundred and sixty-six years later, Othenio Abel went to Kremsmünster Abbey to help Father Leonard Angerer catalog the fossils stored there. In the more than two centuries since the Jesuit fathers had gathered the remains, several more mammoth teeth from the neighborhood had been donated to collection, and the abbey had changed hands from the Jesuits to the Benedictines. Figuring out which tooth came from Laimstetten would have been pure guesswork without Merian's detailed illustration. After comparing the teeth in the collection, Abel found one that, accounting for some damage from handling over the centuries, was a very close match. Angerer was less confident because the tooth in his collection was heavier than the weight given by Merian. The visual match is so close that most later paleontologists have sided with Abel to say the original tooth, a mammoth's tooth, is the one that still resides in the abbey.

As the century rolled on, skepticism about giants grew, but the belief in them never quite died out. At the same time, unicorns were suffering their own devaluation. While skepticism in both the animal and the medicinal substance grew, it was hard to completely abandon the belief, especially when large pieces of ivory, such as the ones Theophrastus had described, still turned up in different parts of Europe.

Otto von Geuricke was not a fool. During his lifetime, he was a philosopher, diplomat, mayor of Magdeburg for thirty-one years, and a respected scientist and inventor. It was for his work the last two capacities that he is probably best remembered. Geuricke invented the vacuum pump and performed public experiments with it that made him an influential member of the European scientific elite. With that resume, it might surprise some to find his name associated with unicorns. His description, published in 1672, is short. In its entirety, it reads:

> It happened in the year 1663 in Quedlinburg, that on the Mountain the common people call Zeunickenberg, where lime is mined, inside the rock a unicorn skeleton was found. The rear portion of the body, as is common in a beast, lay back, head up, but, extending lengthwise from the brow was a horn, the thickness of a human leg, and so in proportion to the length of almost five cubits. Primarily through ignorance,

> the skeleton of the animal was broken and extracted in pieces. Together with the head with the horn and some ribs, spine, and bones, were given to the Reverend Princess Abbess of the place.

The passage gives no indication that Geuricke saw the bones himself, though he had plenty of opportunity to do so. Quedlinburg is less than thirty miles from Magdeburg, where he was mayor, and much of Guricke's technical innovation was aimed at making mining safer and more efficient in places like Zeunickenberg. Guericke was an important enough scientist that his books were read and discussed all over Europe. Several of his peers, though not a great number, noticed this odd entry in his book. The Quedlinburg unicorn was mentioned a few times over the next decades. One of the most interesting retellings of the story appeared in a catalog and commentary on the great collections of Europe published in 1704 and reissued in 1714. Next to a short discussion of *unicornu fossile*, bemoaning the recent sharp drop in prices, the editor, Michael Bernard Valentini, included an illustration of three types of unicorns, *unicornu fictitium*, the fictitious unicorn, with a picture of the classic horse with a horn; *unicornu marinum*, the sea unicorn, with a picture of a narwhal; and *unicornu fossile*, with a picture of the Quedlinburg skeleton. The illustration is bizarre. There's no other word for it. It is a side view of a skeleton with a horse-like skull bearing a long, straight horn. The spine of the animal is dead straight and follows the line of the horn. Only the front legs are present. They have hooves. The spine has some ribs and ends with a ring-shaped bone and three, small curved up bones that might indicate a tail. Valentini credited the drawing to Johann Mäyern, the astronomer of Quedlinburg. Mäyern (or Mayer or Meyer) published a popular almanac, but the illustration is not in any of the issues from those years (he died two years after the discovery). It is most likely that he published the story and illustration as a single-page broadside that has not survived. Valentini, repeating the story, created a small of amout of interest in it. A few travelers mention dropping by the abbey to hunt for the bones. Unfortunately, they report, such curiosities were all kept unlabled in a store room. By then, the Abbess and other witnesses had all died. What they could confirm was that bones and ivory were regularly discovered in the mines and nearby caves.

Mäyern's drawing might have dropped into obscurity had not a much better version of the *unicornu fossile* illustration appeared in a well-studied scientific work by the great Gottfried Wilhelm von Leibniz. Leibniz is probably best known for being the co-inventor, along with Newton, of calculus but, like Geuricke, he had wide ranging interests and made important contributions to a number of disciplines. He wrote about philosophy, medicine, physics, linguistics, history, and politics. He tinkered with lamps, clocks, and pumps, and invented an adding machine. He also took a shot at geology and paleontology. In 1690, his patron, the elector of Hanover, commissioned him to write a history of the House of Brunswick. Leibniz chose to start with the geological prehistory of the land as a background for the human and dynastic history. That section of the history is the only part he completed. A large part of it dealt with fossils. Leibniz cataloged and analyzed the fossil shells in his region. Following that, he looked at some of the other difficult organic remains buried in the mountains. In discussing ivory found in the earth, which he had no problem believing was real, organic ivory, he gave a full section to retelling Geuricke's story of the discovery. The plate prepared to illustrate it gave half of the page to Mäyern's skeleton with the caption "Image of a skeleton excavated near Quedlinburg." The other half of the page shows the tooth of a mammoth, which he captioned "Tooth of a marine animal unearthed from a hill of clay at Tidae, near Stederburg."

Leibniz wrote the geology in 1691-3 and it was not published during his lifetime. Shortly before his death in 1716, the Elector asked him to prepare that work for publication and assigned an engraver, Nikolaus Seeländer, to prepare some illustrations. Seeländer's version was made public in 1749 when the then librarian of the house of Brunswick edited the treatise into chapters and published it in Latin and German as *Protogaea, or A Dissertation on the Original Aspect of the Earth and the Vestiges of Its Very Ancient History in the Monuments of Nature*. Seeländer's illustration is more detailed and carefully executed than Valentini's, which has a slightly cartoonish feel about it. Seeländer uses dotted lines to indicate missing vertebrae. The sources for most of Seeländer's illustrations are known. In all cases, they are very faithful reproductions of the originals. He was not one to incorporate flights of fancy, so we can be confident that his version is the more accurate copy of Mäyern's.

What was it that the Zeunickenberg miners brought out and showed to Mäyern and the Princess Abbess? After the publication of *Protogaea*, several more visitors took a shot at figuring out what it was. More than one thought the illustration and bone fragments resembled those of a rhinoceros, though they were at a loss to explain what a rhinoceros would be doing in Germany. In 1866, when Oscar Fraas included the drawing and story in his *Vor der Südfluth!* (Before the Deluge), a history of the prehistoric world. Fraas commented that the drawing appeared to be a composite of bones from more than one animal, probably a horse with a mammoth tusk attached to the forehead. Of course, Othenio Abel took an interest in the Quedlinburg unicorn. He had already casually mentioned the story in several of his books, crediting Fraas as his source, before he decided to get serious and figure out what the skeleton really had been in 1925. The actual bones had long since disappeared and no other drawings or descriptions had ever been made of them. The best evidence Abel had to work with was Seeländer's plate. Like Fraas, he immediately recognized that it was not a single skeleton; the bones came from at least two individuals and two different species. The skull is that of a woolly rhinoceros. The teeth, scapulae, and vertebrae are from a mammoth. Most of the spine has been reassembled backwards and upside down. What at first glance look like ribs are actually the dorsal spines that are part of the individual vertebrae. These spines are what form the characteristic hump over a mammoth's shoulders. The loop at the bottom of the spine is the first cervical vertebra turned sideways. And the horn; what is the horn? It's too long to be a walrus tusk and too wide to be a narwhal tooth, Leibniz's preferred explanation for fossil ivory. Rhinoceros horns are not made of bone or ivory. They're made of keratin, the same material as hair and finger nails. It's unlikely that the learned burgers of Quedlinburg would have mistaken that for a unicorn horn. That left mammoth tusk, which easily meets the length and width requirements of the description. It takes a little more speculation to explain its being straight and not curved. There are two possibilities here. One is that the tusk was badly enough broken up that the people who reassembled it had the freedom to make it any shape they wanted. The other possibility is that it came from a different kind of extinct elephant, such as the straight tusked elephant, *Palaeoloxodon antiquus*, a species that went extinct about 25,000 years before the mammoth, but whose bones are as common in that part of Germany as mammoths' are.

Though the attempts to explain the Krems and Quedlinburg discoveries leave much to be desired from our perspective, they were an improvement from the time of Habicot and Riolan. While Habicot stubbornly hung on to his preferred explanation that the bones came from a giant, Riolan tried out three explanations without ever settling on one. The first was that the bones came from a giant; he rejected this one out of hand. The second was that the bones came from some rare, but known, animal; he mentioned elephants and whales because they were the biggest animals they knew of, not because he had any knowledge of their skeletons. The third possible explanation was once more that they weren't bones at all; they were jokes of nature, mineral constructions that mimicked the appearance of bones. By mid-century, it was the third explanation far less frequently invoked. It hadn't completely been abandoned, but it had become the last choice for explaining mysterious bone. No one questioned the organic origin of the Krems and Quedlinburg discoveries. Giants, dragons, and unicorns, while fanciful, were all living things. What was missing was the knowledge necessary to figure out exactly what kind of animal produced a given set of giant bones.

By the last quarter of the century, traveling elephant shows had become common enough that they penetrated the remoter parts of Europe. In June 1681, a showman named Wilkins brought a young Asian elephant to Dublin, Ireland and set up a booth near the custom house to show it. Early on the morning of Friday the seventeenth, the booth caught fire and the poor creature was killed before he could bring it to safety. Wilkins realized there was still money to be made if could salvage the skeleton to show. He was able to have a troop of musketeers sent over to guard the corpse from souvenir seekers while he set out to hire as many butchers as he could to clean the bones before the rotting flesh became a public nuisance. Late in the day, a doctor named Alan Mullen heard about the elephant and rushed over to negotiate with Wilkins. Mullen wanted to have an orderly dissection with artists ready to make renderings of each part. Wilkins was willing to let Mullen direct the work of the butchers, but insisted that they finish the work in one day and dispose of the smelly parts before Sunday when they would not be allowed to work. Mullen ordered the butchers to start working immediately. They worked through the night and into the next day. Mullen wrote up descriptions and measurements of the elephant's parts and sent them to Will Petty of the Royal Philosophical Society in London. In his report, he expresses disappointment that he hadn't been able to do a more thorough job. He needn't have been so humble. His examination was far superior to anything that been published in Europe (in India, veterinary treatises on elephants had been available for centuries). Petty had Mullen's letter published as a pamphlet along with a second letter from Mullen on the structure of the eye. In the forty-two pages dedicated to the elephant, he describes all of the major organs and some of the muscle groups, but gives surprisingly little space to the bones. This lack is made for by a trifold diagram of the reconstructed skeleton which Wilkins had managed to assemble and put on display. A second illustration was dedicated to just the skull. Habicot and Riolan had both had to admit that they had never seen an elephant and regretted the lack of materials that would allow them to make a proper comparative study of the bones. Thanks to Wilkins, Mullen, Petty, and the poor nameless elephant, the savants of Europe finally had something to use for comparison when examining giants' bones.

CHAPTER 2

THE IVORY TRAIL

If the giants and unicorns of France and Germany had been the only mammoths that Western European thinkers in the late seventeenth century had had to work with, it would have been decades, possibly even a whole century, before they recognized them as a species separate from modern elephants. As it was, most of the eighteenth century would pass before they realized African and Asian elephants were more than one species. Once the bones were properly identified as elephantine in nature, there would have been no mystery about them that needed to be solved. Normal elephants north of the Alps could easily be explained by Hannibal, the Romans, or by the stories of medieval kings receiving elephants as gifts from eastern emperors, such as Charlemagne's elephant. It was the arrival of Siberian ivory called "mammoth" that created a distinct problem. This ivory looked like elephant ivory, but everyone knew elephants could not have lived in the frigid climes of Northern Asia and there was no known history that had someone bringing large numbers of them there. There were no Roman armies or powerful monarchs in the far reaches of the taiga and no good reason for them to go there. For the idea of the mammoth as an entirely separate being to get to Europe and require explanation, we need to look at the trade that brought the ivory there.

At least eight hundred years before the word "mammoth" arrived in Western Europe, ivory from the lands that would become northern Russia was being traded on the international market. Sometime before the year 890, a Viking named Ohthere showed up at the court of King Alfred of Wessex. Later generations would call the king of Wessex Alfred the Great, the unifier of England and the man who let a peasant woman beat him for burning her bread. That last part is one of those apocryphal stories of a kind king traveling incognito to learn about the hard lives of the common folk. Even though the bread story never happened, Alfred was a pretty impressive guy. Along with unifying the petty English kingdoms, he beat back Viking colonizers, made London the capital and the most important city in England, and took time off to translate important works of literature into the common language. While at the court, such as it was, Ohthere presented Alfred with a large piece of ivory and described his travels in distant lands (a story that did not involve plundering England). Alfred was pleased enough that he wrote down Ohthere's story and attached it to his translation of the world history written by the early Christian writer Paulus Orosius, the same historian who described the war with Theutobochus. There is no record of what he did with the ivory.

The story Ohthere told Alfred began with an account of his wealth and his lands before moving on to his travels. Ohthere was a lord in Hålogaland, the northernmost settlement of the Norwegians, as he described it. He told the king that he had "wished to discover on some occasion how long the land lay due north, and whether any man lived due north of the wilderness." Ohthere described sailing north for three days, east for three days, and finally south for five days. There he stopped at the mouth of a great river. This voyage would have taken him around North Cape and the Kola Peninsula and into the White Sea. At this point in his story he admits, "Chiefly he went thither, to increase the examination of the land, because of the horse whales [walruses], because they have a very noble bone on their teeth . . . and their hide is very good for ship-rope." What made Ohthere's hunting trip interesting to Alfred was that there was no record of anyone having sailed into those waters before. The very fact that the land turned east and didn't continue north to the pole was news to the English. Ohthere's tale is also the first recorded mention of the walrus.

But Ohthere didn't just give Alfred some nice walrus ivory; he gave him ivory from a completely unknown animal. No wonder the king was impressed. It's important to note that Ohthere didn't sail north on a whim; he had a specific economic goal in mind. He was familiar with walrus ivory and knew where to look for it. Before his voyage, Ohthere's access to ivory would have been through overland trade with Sami (Lapp) middlemen. As the lord of Hålogaland, the local Sami reindeer herders paid him an annual tribute. While in the White Sea, Ohthere spent considerable time interviewing leaders of the people he called Bjarmians and learning about conditions in the region. By proving a sea voyage to the White Sea was possible, he eliminated the middlemen. Others followed.

The extent of Norse trade in the White Sea is hard to estimate. Most of the Norse records of that era are sagas, whose authors were more interested in recording adventures, feats of arms, and impressive plundering than in documenting mundane business transactions. But, buried in those sagas are plenty of indirect hints that regular, peaceful trade was happening. Many of the plundering expeditions focused on the same spot, the mouth of the Dvina River, which was the northwestern end of several river trading networks and the place where Ivan the Terrible would later order the town of Archangel to be built. The sagas of St. Olaf make clear the importance if this spot. Sometime in the early eleventh century, Olaf and Karli of Hålogaland entered into a partnership for a trading trip to the Dvina mouth. Along the way, they were joined by Thorir the Hound. After conducting some profitable business, they began to leave but, on the way home, Thorir convinced the others to return and plunder the region. The main point of the story was to set the background for later conflicts between Olaf and Thorir leading up to Olaf becoming a saint. However, in passing the saga lets us know there was a regular seasonal market at the Dvina mouth governed by an established truce. Even though they returned and proceeded to plunder the Bjarmians, Olaf, Karli, and Thorir waited for the truce to expire before doing so. There were rules for trade in the area that even the bloodiest Vikings respected.

The Norse trade in the White Sea declined around 1250. In that century, the northern hemisphere was beginning to cool. Expanding sea ice near North Cape made the voyage increasingly difficult. At the same time, the Norse faced strong competition from merchants on the Volga who paid good prices for walrus ivory to sell in Persia and the Middle East. Even if the prices weren't that great, dealing with mostly honest river traders must have been preferable for the Bjarmians to selling to people like Thorir who were going to come back later, kill everyone, and take their money back. Leaving the White Sea did not put the Norse out of the ivory business. During the same years that they developed a market in Western Europe for walrus ivory, Norse sailors ventured out into the Atlantic and colonized Iceland and Greenland. When these colonies were brought under the control of the Danish-Norwegian crown, the preferred way for these distant communities to pay their taxes was to send walrus ivory. The art historian Paul Williamson has estimated that most of the ivory carved in Europe from 1000 to 1300 was walrus ivory.

In the century following Ohthere's voyages, Muslim writers began to speak of walrus ivory as a product of the north. Though there are some fascinating, but cryptic, references before the tenth century, the first unmistakable mention of Arctic ivory in the Middle East comes from the geographer Muhammad ibn Ahmad al-Muqaddasi. In 985 he wrote that the products of the north were "sables, miniver, ermines, and the fur of steppe foxes, martens, foxes, beavers, spotted hares, and goats; also wax, arrows, birch bark, high fur caps, fish glue, fish-teeth, castoreum, amber, prepared horse hides, honey, hazel nuts, falcons, swords, armor, khalanj wood, Slavonic slaves, sheep and cattle. All these came from Bulgar. . . ." Bulgar or Great Bulgaria, on the middle Volga River, was a major market trading manufactured goods from the Islamic world for forest products from the north. The term "fish-teeth" in the middle of al-Muqaddasi's list is the primary name by which walrus ivory was known when traded into southern and eastern Asia. Fifty or so years after al-Muqaddasi, Abu Rayhan al-Biruni wrote that "the Bulgar bring from the northern sea teeth of a fish over a cubit long. White knife hafts are sawed out of them for the cutlers." He goes on describe how no part of the ivory was wasted. Al-Biruni and several other writers following him commented on the fantastic prices paid for fish-teeth in Arabia and Egypt, despite the fact that both of these places had access to Indian and African elephant ivory. In time, fish-tooth knife and sword handles became so popular in South Asia that they were even imported into India for exorbitant prices.

For our purposes, the most interesting comment by a Middle Eastern writer comes from Abu Hamid al-Gharnati in the mid-twelfth century. Al-Gharnati was born in Andalusia, Spain and traveled the length and breadth of the Muslim world. He lived for several years in Bulgar. In writing about his years there he describes the walrus ivory trade, and adds that "teeth were also found in the ground like elephant's tusks, white like snow, one weighing two hundred menn [250 pounds]; it was not known from what animal it was derived; it was wrought like ivory, but was stronger than the latter and unbreakable." He goes on to say that these teeth could be sold in Khorezm (in modern Uzbekistan) for a great price. "Elephant tusks" dug out of the ground in northern Asia can only mean mammoth ivory. But, we should be careful about reading too much into this statement. Mammoth ivory often erodes out of the banks of the Volga and Don in the vicinity of Bulgar. That the tusks were being sold in Bulgar is not necessarily evidence that Siberian mammoth ivory had entered trade networks going south and west at that time. It is, however, evidence that, when it showed up, ivory merchants viewed it as just as desirable as other forms of ivory.

Along with west and southwest, the third direction a regular trade in Siberian mammoth ivory could have developed was to the southeast—China. Since the early nineteenth century, naturalists and historians have been fascinated by the idea that mammoth ivory was known in ancient China. Formal trade between Russia and China began in the 1690s when literate Westerners arrived in Siberia in meaningful numbers. Twenty years later, they discovered that the local fur traders had already developed a regular trade in mammoth ivory with China. The Chinese have known and loved ivory since the earliest times. There is no record on either the Russian or Chinese side that the Chinese showed the slightest resistance to accepting mammoth ivory as real ivory. Does this mean they were already familiar with it? It might. By a lucky coincidence, at the same time Russian

Cossacks started hunting for mammoth ivory to sell to China, the Kangxi emperor, who reigned from 1661 to 1722, was writing a natural history of his own empire. When Western naturalists became curious about Chinese knowledge of the mammoth in the early nineteenth century, they discovered that all the relevant documents had already been collected for them by the late emperor.

Some otherwise unidentifiable animals appear in ancient Chinese literature that could be mammoths. The best candidates are usually referred to as giant rodents (*shu*). Mammoths might seem unusual candidates for inclusion in the order rodentia, but there is a certain logic to it. An uneducated person finding a mammoth carcass buried in the ground would try to compare it to an animal they knew. Familiar large animals—cattle, horeses, bears—do not burrow. Most medium-sized burrowing mammals are clearly what they are, even in skeletal form—rabbits, badgers, martens. But burrowing rodents come in hundreds of types and sizes, from the very tiny to the big enough to eat. Lacking a specific name for a new subterranean animal, it wouldn't be unreasonable to simply call it a rodent. In the *Shen i king*, a book by So Tung-fang, minister to the early Han dynasty emperor Wu (140–87 BC), the following passage appears: "In the regions of the north, where ice is piled up over a stretch of country ten thousand miles long and reaches a thickness of a thousand feet, there is a rodent, called *k'i shu*, living beneath the ice in the interior of the earth. In shape it is like a rodent, and subsists on herbs and trees. Its flesh weighs a thousand pounds and may be used as dried meat for food; it is eaten to cool the body [i.e. reduce fever]" T'ao Hung-king, in the fifth century, wrote a pharmacopeia entitled *Ming i pieh tu* and, in the eighth century, Ch'en Ts'ang-k'i wrote a work called *Pen-tsao Shi-I* on the omissions of previous pharmacopeias. Both included sections on the *fen* or *fyn* (an animal that moves in the ground) also known as *fyn shu* (hidden rodent). They told their readers that there were two types of fen, the common small mole and the other fen, which was the size of a water buffalo. Later writers right up to the time of the Kangxi emperor repeated these stories.

The Kangxi emperor believed that the description of deep ice in the *Shen i king* showed accurate knowledge of the Arctic Ocean and he made the connection between the various shu and the mammoth. He even made a connection between the mammoth and the elephant. He confidently wrote, "In both these points, the ancient books are confirmed . . . In Russia, near the shores of the northern ocean, there is a rodent similar to an elephant, which makes its way underground and which expires the very moment it is exposed to light or air. Its bones resemble ivory, and are used by the natives in manufacturing cups, platters, combs, and pins. Objects like these we ourselves have seen." The emperor didn't need to trust Westerners for confirmation of this conclusion. Before he completed or published his opus, Tulishen, a trusted agent of his, returned from a mission to the shores of the Caspian Sea and reported of Russia: "In the coldest parts of this northern country is found a species of animal which burrows under the ground, and which dies when exposed to the sun and air. . . . The Russians collect the bones of this animal, in order to make cups, saucers, combs, and other small articles . . . The foreign name of this animal is *mo-men-to-wa* [mammoth]; we call it *k'i shu*." A giant mole that lives in the frozen north and is never seen alive certainly sounds like a frozen mammoth carcass. The emperor and his agents believed they were one and the same. It's tempting to leave it there, but there is a problem in accepting the emperor's conclusion that the *k'i shu* was a mammoth. None of the old sources mention ivory. In China, ivory isn't just a beloved artistic medium, as in Europe, it also had medicinal uses. The pharmacopeias mention the medicinal value of its meat, but never mention ivory.

Whether the early Chinese actually traveled to Siberia to buy ivory is a difficult question to answer. There are no surviving accounts of Chinese travelers in the North or even second-hand references of such accounts. But they did have some knowledge of the people to the north. The chronicles of the Eastern Zhou (770–256 BCE) mention the tribes in the Amur valley. Sources from the Han dynasty (202 BCE–220 CE) describe tribes farther up the Pacific coast possibly as far as the Bering Straits. Walrus ivory artifacts from that period have been found on both the Siberian and the Alaskan sides of the Bering Sea that appear to have been carved with metal tools—tools that could only have come from China. There is good evidence that the tools were paid for with ivory. Beginning in the sixth century BCE, Chinese chroniclers wrote about a people called the Sushen who lived north of China in what are now Manchuria and the Russian Maritime Province.

The chronicles describe the Sushen as difficult and warlike and take time to describe their weapons, which included armor made of bone. In more recent times, anthropologists have described people on both sides of the Bering Sea making armor out of walrus ivory.

Sometime during the Tang dynasty (618–906), a new substance began to appear in China. It was called *ku-tu* and was a form of tribute from northern tribes. Hung Hao, a twelfth century diplomat, described it as: "The *ku-tu* horn is not very large. It is veined like ivory, and is yellow in color. It is made into sword-hilts." Sien-yu Ch'u, a poet and calligrapher wrote that "*ku-tu* is a horn of the earth." This could describe mammoth, walrus, or even narwhal ivory, the latter two often being collected along the shore rather than hunted. The word *ku-tu* spread to Central Asia, Persia, and the Arab world picking up various spellings and pronunciations along the way. Although it might not always have referred to the same material, it always had the same approximate meaning of a substance of unknown origin, brought by northern merchants, and very desirable for making knife and sword hilts.

In the early twentieth century, the American anthropologist Berthold Laufer wrote several papers on ivory in China in which he looked at the trade in walrus ivory. His conclusion was that *ku-tu* was almost entirely walrus ivory. As for that "almost," he explained that mammoth, walrus, and narwhal were collected by people of the north and traded south. Along the way, they would be cut to sizes that were easy to transport, with the bad parts cut away, and indiscriminately mixed together. Once away from the coast, the next merchants down the line would have had no point of reference to understand what any of those animals were. It was all *ku-tu* to them. By the time the ivory made it to China or Central Asia no one could say what kind of animal produced *ku-tu*.

By the early modern period, long-distance trade routes had penetrated Siberia on all sides and, on all sides, end consumers wanted ivory. For centuries, most of that trade consisted of walrus ivory with occasional pieces of narwhal and mammoth mixed in. In the late medieval period, Russian merchants pushed farther east and, in doing so, eventually came into direct contact with people selling giant tusks, much too big to have come from a walrus. The people selling them called them "mammoth."

In the thirteenth century, as the Norsemen abandoned the White Sea region, it came under the political control of the Russian city-state of Novgorod, located near the Baltic Sea south of modern St. Petersburg. Novgorodian merchants had been trading on the White Sea almost as long as the Norsemen, the most important difference being that, while the Norse were only visitors to the area, the Russians slowly built up a permanent presence by setting up trading posts and controlling important portages between river systems. The merchants were followed by settlers who established farms and towns. Many of them came fleeing the Mongol invasion that conquered all the other Russian states except Novgorod and its neighbor Pskov. All of these settlers eventually became subjects of Novgorod. In 1251, Norway and Novgorod concluded a treaty ceding the entire White Sea to to the Russian city-state. By then, the more adventurous merchant families had advanced out of the White Sea watershed eastward to the Pechora River watershed covering the west side of the northern Ural Mountains. Though they were never able to exert control across the mountains into Siberia, they were able to establish regular trade there. This created a direct path from the mammoth regions of northern Siberia to Novgorod and Western Europe.

The Russians called the area straddling the northern Urals "Yugria." The Yugrians were two different peoples, the Samoyeds (now Nenets) along the coast and the Voguls (now Mansi) along the rivers. Today, the last Mansi only live on the eastern side of the mountains along the lower reaches of the Ob River. When the first Novgorodian merchants arrived, the Mansi populated a large part of the Pechora basin on the western side of the mountains. The Mansi are one of the keys to explain how the idea of a mammoth came west. The most recent linguistic research argues that the word "mammoth" is almost certainly derived from a Mansi construction meaning "earth horn." The etymology is important. It shows that the Mansi were not just the middlemen connecting the Novgorodians with ivory collectors farther east. It shows that the Mansi had clear—probably first-hand—knowledge of the origin of mammoth ivory. That the Russians picked up the word indicates that, at the point when they began to distinguish between mammoth and walrus ivory, it was Mansi merchants who they turned to to provide a name for the less familiar ivory. It might have been as simple a matter as someone saying, "Hey, why are these two pieces different?" It might also have been that larger pieces of mammoth ivory were being traded, even whole tusks. This was definitely happening by the beginning of the seventeenth century. To best understand the slow process of differentiating the mammoth ivory from that of the walrus, let's go to the end of the story and work our way backwards.

On the July 16, 1618, a British diplomatic mission, led by Sir Dudley Digges, arrived in Archangel, the primary Russian port on the White Sea. The purpose of the mission was to negotiate a loan for the tsar, Michael Romanov. Michael was in no mood to negotiate. When the advance party of the English arrived with half of the funds, he seized the money and sent them on their way. On hearing this, the ambassador ordered the rest of the mission to turn around and return home. By then it was October and Archangel was iced in. Sir Dudley took a small party, and the rest of the money, and traveled overland to Swedish territory on the Baltic Sea, and, from there, home. The rest of the party returned to Archangel to wait out the winter and the first ships of spring. Among those was the mission's chaplain, Richard James. To occupy his time during the long winter nights, James compiled a lexicon of new and unusual words. Three quarters of the way through his lexicon, we find this entry: "maimanto, as they say, a sea Elephant, which is never seene, but according to the Samites [Samoyeds], he workes himself under grownde and so they find his teeth or homes or bones in Pechare [Pechora] and Nova Zemla." James's *maimanto* is not a walrus. Although sea elephant was one of many names used to describe walruses, James had already recorded the Russian word for walrus (*mors*). He also included the important details that mammoths were believed to live underground and were never seen alive, neither of which is true for walruses.

Eight years before James wrote about the *maimanto*, Josias Logan visited Pustozersk on the Pechora River at the other end of the trade route to Archangel. Logan was the first representative of the British Muscovy Company to be sent to the town. Pustozersk sat on the flanks of the Ural

Mountains, and by establishing a post there, the company hoped to acquire the wealth of Asia and eliminate most of the middlemen between England and China. His first sight of the town could not have been encouraging. After a difficult voyage across the Arctic coast of Russia, Logan arrived in July 1611 to find half the town in ashes and the governor missing. The only representatives of the government with any authority were an informal committee of tax collectors who were not sure they trusted him. While he waited for their approval to stay, Logan visited the local market and purchased "a piece of an Elephants Tooth," which he sent back to London with an optimistic letter claiming that this must be evidence of an easy road from Pustozersk to China. At the time, Westerners had no idea how deep China reached into Asia. Many thought it was right over the Urals. Richard Finch, who was with Logan that day in the market, also wrote to the company about the transaction: "there is in the Winter time to bee had among the Samoyeds, Elephants teeth, which they sell in pieces according as they get it, and not by weight. . . . It is called in Russe, Mamanta Kaost." Finch's letter is especially important. There is no way to mistake what he means: the Russians had elephant tusks and they called them mammoth bones. Finch's account was published in 1625 making it the earliest known appearance of any form of the word "mammoth" in print. The letter itself is the earliest appearance of the word in any form in a Western European language. There are, however, older written, but not formally published, Russian sources that use the word.

The earliest known record of some form of the word "mammoth" is in the account books for the year 1578 at the Anthony of Siya Monastery near Archangel. The record is two simple words "pyatye mamantovakos," or "one fifth mamant bone." We can tease quite a bit of information out of those two words. The first few facts are pretty obvious: by 1578, the word "mammoth" was known west of the Urals, mammoth ivory was being traded across the Urals, and the Russians receiving the ivory knew that it was different from other forms of ivory, particularly the walrus ivory that they would have already been familiar with. That's a lot. Beyond that, we can deduce that mammoth ivory was not new to them at that time. The words come with no explanation from the writer to the reader. The ivory is not a marvelous treasure; it's simply something in their inventory. They were familiar enough with mammoth ivory to recognize that this piece is one fifth of a tusk, not more, not less. This familiarity means that a regular trade in mammoth ivory, recognized as such, had been going on for some time, at least most of the life of the writer and his intended readers. This pushes the trade back to the 1550s, if not further. How much earlier can we push it? That's a tricky question. Now we have to turn to maps.

Item MS 24065 in the collections of the British Library is a large, hand-drawn and painted map of the world. It is signed by Pierre Desceliers and dated 1550. The map bears the coats of arms of King Henri II of France and the duc de Montmorency and was most likely custom-made as a gift to the king. This style of map is meant to be viewed spread out on a table rather than hung on a wall. Half of the text is legible to someone standing on the south side of the map and half by someone on the north side with the centerline lying at about eleven degrees north. The map is decorated with colorful vignettes of animals and people in distant parts of the world. There are twenty-five text boxes. Standing in Northwest Russia, near the White Sea and Scandinavia is an elephant. It stands at the end of the Siberia to Pechora to Archangel ivory trail. Desceliers intentionally put the elephant in that spot. All the other animals and people on the map are where they would be in life; camels are in Arabia and polar bears are in Greenland. The monsters are where legends say they should be; there is a gryphon in Central Asia. An army of Amazons is on the march where Herodotus said they should be. Any doubt that the elephant belongs in that spot is dispelled by a text box which tells us that the trade products of the Rucheni (Ruthenians or Russians) are "valuable pelts, falcons, gyrfalcons, white elephants [*ylefanz blanc*], bears, moose and others that they carry to other parts of the world." With the exception of the elephant, Descelier's description is an abridgement of a passage from an edition of Marco Polo well-known in his day. Three years later, Desceliers would produce another world map with an elephant in the same place. Although this one was without a caption, it demonstrates that the first was not an anomaly or mistake. He intended for that elephant to be there.

Is Desceliers's Russian elephant an indication that the mammoth ivory trade had reached the White Sea and that Western mapmakers knew about it by 1550? The most honest answer is somewhere between "maybe" and "probably." Desceliers left us no clues. Many mapmakers of the time very scrupulously let us know their sources of information. Others just as scrupulously kept them secret, often because they had bribed ships' captains to divulge information that was considered a state secret by some countries. Desceliers was one of the latter. We don't know what his sources were or what they told him. Desceliers was part of a group of mapmakers called the Dieppe school because they lived and worked in the vicinity of Dieppe on the north coast of France. King Henri II was a major patron of the mapmakers. On two unsigned, earlier maps from the school (Desceliers might have been the creator of one) a different, unnamed beast inhabits the space where Desceliers put his elephant. Both animals are gray and have thick bodies and tusks jutting up from their lower jaws, somewhat like a boar's tusks. On the 1546 map, today known as the Dauphine map, this animal has a body like a domestic pig: thick, rounded, longish, and with thinner bent legs and cleft hooves. The other map from the 1547 Vallard atlas (named for an owner, not the artist) has a more elephant-like body, shorter in length and less rounded, with long, thick, and straight legs. Desceliers is often credited with having produced the pig-like one. Neither of these animals is a modern elephant, but they don't resemble any other known species. They do, however, bear a resemblance to an animal on an earlier map by one of the most famous mapmakers of the century.

Martin Waldseemüller is best known for his 1507 map of the world that was the first to use the word "America." With such a claim to fame, it's not surprising that first thing most writers say about his next world map, the *Carta Marina* of 1516, is that it does not use the word "America." What they should have been saying was, "is that an elephant next to Norway?" The animal on Waldseemüller's map looks enough like the animal on the Vallard map that it's safe to say that it was the source for it. Waldseemüller's animal has the same elephant-like, blocky body and thick legs. It also has the fan-like ears of an African elephant. What it lacks is an elephant-like head. It has no trunk and its tusks jut up from its lower jaw. Unlike the two Dieppe maps, Waldseemüller was good enough to provide a caption to tell us what the animal is supposed to be: "The walrus [morsus] is an elephant sized animal with two long, quadrangular teeth. It is hindered by a lack of joints. The animal is found on promontories in Northern Norway where it moves in great herds." His morsus appears on a half-dozen other maps by different mapmakers over the next twenty years.

Waldseemüller not only brings us back to the walrus; he demonstrates a crisis of knowledge that was happening during the Renaissance. It was a crisis that, in part, fueled the Renaissance and led to the scientific revolution. During the Renaissance, the world the Europeans knew became vastly larger in size and complexity. In the south, Vasco da Gama demonstrated that the Atlantic and Indian Oceans were connected and that is was possible to sail from Europe to the rich markets of Asia. Columbus demonstrated that exotic markets were reachable by sailing west. A few years later, his peers realized that these lands were not part of Asia, but entirely new and unsuspected lands. Economic ambition, the search political advantage, and plain curiosity drove Europeans to explore this new larger world. For the next two centuries, it seemed every ship brought back tales of new places, cultures, plants, and animals. As literate Europeans tried to make sense of these reports, they found old organizations of knowledge insufficient to the task. One of those organizations was the bestiary. A bestiary was a compendium of known animals. Many were beautifully illustrated books combining scientific knowledge of ancient authorities, like Aristotle and Pliny, with moral lessons from the church. The walrus was unknown to the ancients and not part of the standard corpus of bestiary animals. Although the ivory had been known for over six hundred years, no illustrations of a walrus are known to have existed before 1500.

The one written source that described walruses and was available to Waldseemüller was the eccentric and prolific genius St. Albert the Great (Albertus Magnus). In the mid-thirteenth century, Albert wrote a commentary on Aristotle's *De Animalibius.* Departing from the usual bestiary formula of moral lessons, his was primarily a scientific tract and included many animals not known to Aristotle. In his section on whales, he says that one type of whale has two long canine teeth, usually one cubit, but sometimes two or even four cubits long. He says that the texture of the teeth is like that of an elephant or a boar. It uses these teeth for fighting and to pull itself up on the rocks where it sleeps. One cubit—eighteen inches—is a realistic size for a walrus tusk. Three to six feet is unknown, but easily within the range for a narwhal tooth or mammoth tusk. Albert's physical description is followed by a bizarre description of walrus hunting that involved sneaking up on them while they slept on the rocks, slipping a rope through a cut in their skin, tying the end of the rope to a tree, and waking them by throwing rocks at their heads, upon which the frightened walruses would run out of their own skins in their rush to get back to the sea. Someone was pulling Albert's leg, so it's safest to assume that the giant tusks were another exaggeration and not evidence of accurate knowledge of mammoths. What's important here is that Albert's prestige—him eventually being a saint and all—led to his story being uncritically believed until the middle of the sixteenth century—the same time that Desceliers produced his white elephant map.

One of the most famous depictions of the walrus came from Olaus Magnus, the last Catholic bishop of Uppsalla, Sweden. In 1518, Magnus traveled into the north of Sweden selling papal indulgences in order to raise money for the construction of a new St. Peters in Rome. He wasn't a very good salesman; he spent most of his time collecting information on native folkways and natural history. In 1539, he published the first part of the results of his research in the form of a lavishly illustrated map. In 1555, he published a text version of his researches carefully linked to the illustrations on the map. Near the White Sea is a monster, with thick short legs, tusks in its lower jaw, spiky hair, and a very annoyed expression at a group of Sami who are throwing snowballs at it. The description in his book begins "To the far North, on the coast of Norway, there lives a mighty fish, as big as an elephant, called morse or rosmari . . ." This hits all the important points used by Waldseemüller. Further on, he writes: "They will raise themselves with their Teeth as by Ladders to the very tops of Rocks, that they may feed on the Dewie Grasse, or fresh Water, and role themselves in it." This matches elements from Albert. He also includes the bizarre hunting anecdote.

Most of Magnus's sources are known to us. Albert's tale was well known in the sixteenth century and was quoted in a half dozen books printed before 1550. During his travels, Magnus met with Bishop Erik Valkendorf of Trondheim, Norway, who had seen walruses during a trip to the northern part of his diocese in 1512 and described them in a letter. The spiky hair on his walrus could have come from Albert or from Valkendorf. What we lack is a source for the legs and the comparison to an elephant. In his description, Magnus names two earlier travelers in Russia who wrote about the importance of the ivory trade with the Middle East, Paulus Jovius and Maciej z Miechowa. Both of these writers mention Albert's hunting tale, neither mentions legs. While it's possible Magnus took those details from Waldseemüller's map or from later copies of it by Laurent Fries, it's unlikely that Waldseemüller simply made up those details. Everything we know about him points to his being a very careful researcher. Albert used the word "elephant" to describe the nature of walrus ivory, but nothing else in his description suggested elephantness. In any case, the tusks in Waldseemüller's drawing are one of the least elephant-like things about it.

These problems point to a lost source known, at least, to Waldseemüller and possibly also to Magnus and one other. Sigismund von Herberstein traveled in Russia on a mission for the Holy Roman Emperor in 1517, the year after Waldseemüller's map was printed, and again in 1526. Herberstein is responsible for the "cz" spelling of "tsar" (i.e. "czar"). Herberstein's original reports are gone, but he published a book-length description of Russia in 1549. To his own observations, he added details gleaned from other travelers' accounts. On the walrus, Herberstein made the same observations on the value of the ivory trade as had been made by Jovius, Miechowa, and Magnus. His physical description adds the detail we've been looking for—legs. He wrote: "It has short feet, like those of a beaver; a chest rather broad and deep compared to the rest of its body; and two tusks in the upper jaw protruding to a considerable length." Short beaver legs are not long mammoth legs, but they are legs and beavers are furry, like mammoths. Long tusks in the upper jaw are more like those of walruses and mammoths than the boar-like lower jaw tusks of Waldseemüller's and Magnus's maps. Herberstein, like Waldsemüller, was a careful researcher. It's unlikely that he would have made up such specific details. Saying that the walrus was a very large animal with legs does not prove that the lost source knew something about mammoths as a source of ivory. The best we can say is that it's suggestive. Herberstein never went beyond Moscow, so he had to have depended on someone else for his knowledge of walruses. If the lost source really existed, was it the same source that convinced Desceliers to change his elephant-like creature (almost certainly inspired by

Waldseemüller) into a true elephant? What did the early sixteenth-century mapmakers and travelers know about the mammoth that other intellectuals of the period did not?

While the mapmakers suggest a lot and leave only questions, the travelers suggest the answer to one of our questions. Maciej Miechowa, mentioned above, was a Polish polymath who wrote the first great history/geography of the lands between Germany and the Caspian Sea, which in those days meant Poland and Russia. *The Two Sarmatias* was published in 1517. Miechowa's description of the walrus is worth looking at in detail.

> In Yugria and Karelia some moderate mountains rise. . . . Beyond the mountains, the ocean is moderate along the whole northern coast: there the fish called the morss [walrus] climbs from the sea onto the mountains by its teeth, chaffing himself during the ascent. And while to the top of the mountain shall come, making a path to be followed henceforth: fall flying to the other side of the hill. The latter nations gather together the teeth which are broad and white and very large and they take the most weight and sell it to the Moscovites: the Moscovites, however, use these to sell to the Tartars and Turks who use them to make handles for knives and swords and also spears.

At first glance, this does not seem to add anything new. The walrus lives by the sea; it climbs ashore using its teeth; the locals trade its ivory to the Russians, who sell it to knife-makers in the Middle East. But it does say something new. Albert said walruses climb onto the shore to sleep. Magnus said the climb to high meadows to graze. Miechowa says they climb high into the mountains and some fall to the other side—that is, inland—where their ivory is gathered, not hunted, by Yugrians. To put it another way, the Yugrians-Mansi—to whom we are indebted for the word "mammoth"—find and collect ivory inland from the Arctic Ocean to sell to the Russians. Miechowa, like so many others for centuries before him and decades after, conflated the walrus and mammoth ivory trades together, but, in his description of the source of that trade, he described only the trade in mammoth ivory. In describing this as the normal form of ivory trade, Miechowa pushes the date of a trade in mammoth ivory, as distinct from walrus ivory, back to the beginning of the sixteenth century. As with the casual way the monks of St. Anthony used the word "mammoth" sixty years later, Miechowa's description of gathering mammoth ivory as a normal thing means it probably dates into the previous century.

Ivory has been a valuable item of trade in all of the civilizations that fringe Eurasia for all of recorded history. When the peoples of northern Eurasia entered the long-distance trade networks that connected them with those civilizations, they met the end-consumers' demand with any type of ivory they had access to. At some point, toward the middle of the last millennium, along a trade route that passed through the lands of the Yugrians/Mansi, mammoth ivory began to move in significant enough quantity that it became distinguished from walrus and narwhal ivory. This occurred sometime before 1500. Still, it took decades for the end-consumers in Europe to absorb that knowledge. Miechowa was given a big hint before 1517 but lacked enough comparative data to know he was hearing something new. Western mapmakers, striving to synthesize an early version of data overload, probably processed and lost one, maybe two, sources that described the mammoth as an animal, not just a source of ivory that was distinct from the walrus. There are probably a lot of lost documents that refer to the mammoth ivory trade hidden in monasteries, chancelleries, and libraries in Russia, the Middle East, and Central Europe. James's "maimanto" was not discovered until 1950, and the St. Anthony monastery account books were unknown until the 1970s. Logan's elephant tooth was only attached to an unknown animal in 1692, and I discovered Finch's "Mamanta Kaost" in 2011. Who knows what remains to be discovered.

CHAPTER 3

EASTERN TRAVELERS

In the years following Logan and Finch visiting Pustozersk and James wintering at Archangel, there must have been many other travelers and merchants who visited Russia and heard about the mysterious ivory-producing animal, the mammoth. Unfortunately, if they left written accounts, we haven't discovered them yet. The next traveler whose thoughts about the mammoth have survived was Nicolaas Witsen, a young member of a Dutch trade mission to Moscow. Witsen's family was deeply involved in the grain trade with Russia, and he had already expressed an interest in the country. The year before the trip he sent a list of questions to the Dutch agent living there about all aspects of life in Russia. After his visit, his casual interest became a lifelong fascination with the Russian far east, the unknown lands beyond the Urals. Witsen didn't keep his love to himself. He wrote, he collected tales from travelers, he encouraged others to go to and learn about that quarter of the world, and he helped those others get published. In doing so, he, more than anyone else, deserves credit for introducing the word and concept of the mammoth to the West.

Witsen came from a prominent Amsterdam family that was active in both politics and the economy. His father was elected the mayor of Amsterdam four times and was also Rembrandt's landlord (in 1658 he caused the artist's bankruptcy by unexpectedly recalling a loan). Born in 1641, he lived during the golden age of the Dutch Republic. During his youth, the Dutch East India Company fielded the largest merchant fleet in the world and controlled the lion's share of trade from Asia. Since his family had a big hand in that trade, he was in a perfect position to learn about the exotic cultures and environments of the rest of the world. Witsen had an extensive education and had already made one trip abroad—to England with his father in 1656—when he was appointed to take part in the mission to Moscow. After taking a roundabout route to avoid cities experiencing an outbreak of the plague, the Dutch party arrived in Moscow on January 20, 1665. It left on May 12. Witsen didn't have an official position in the embassy; it was primarily intended to be an educational trip preparing him to take his position overseeing the family's business interests. This gave him plenty of free time to explore the city, though he often had to slip away from his guards in disguise to do so. During those four months, he met Samoyeds, Tartars, and Persians; became friends with Patriarch Nikon and Archil Bagration, four-time king of the Georgian kingdom of Imereti; toured the markets; made drawings of buildings; and began collecting travelers' accounts of the far east to make a "contribution to the explanation and description of the earth." Many of the contacts he made during that trip would correspond with him for the rest of their lives. One of the most important of those was a distant cousin, Andreas Winius, who at the time served as the mission's interpreter, and would eventually rise to become the minister in charge of the Siberian Prikaz, the office that oversaw many of the royal monopolies on products coming out of Siberia, which, by the end of the century, included mammoth ivory.

The Siberia that the Siberian Prikaz watched over was a wild and unruly place. Few in Moscow had a clear idea of the extent of the territory. Western Europeans had even less of an idea. For them, the area east of the Ural Mountains was one of the least-known parts of the planet. Russia did not necessarily acquire Siberia "in a fit of absentmindedness," as John Seeley once said of the British Empire, but their conquest of the east was nearly that unintentional and chaotic. When Ivan IV (not yet terrible) inherited the throne of the Grand Duchy of Muscovy in 1533, his realm covered slightly less than half of the European part of modern Russia and had dangerous borders in all directions except the ice-bound north. Muscovy was frequently isolated by wars with its neighbors. In the west, Poland and Sweden cut them off from the Baltic Sea and the rest of Europe. In the south and east, the successors of the Tatar Golden Horde, which had once conquered and devastated the Russian principalities, still controlled the rich black-earth lands of the Eurasian steppes. These khanates regularly raided Russian villages for slaves, imposing an enormous cost on Moscow for defense. The Khanate of Kazan, the successor to Great Bulgaria, was closest to Moscow. It straddled the Volga and cut the Mucovites off from the rich fur trapping lands of the southern and central Ural Mountains, which might have helped pay for that massive defense budget.

It's hard to overestimate the importance of the fur trade in the development of the Russian state. From the very beginning, fur—primarily sable fur—was the single most valuable trade commodity for the Russian principalities and the largest source of revenue for the princes. For centuries, very little coined money circulated in the Russian lands; furs functioned as the primary currency. It was the quest for fur trapping country that had led the Novgorodians and other Russians to move out of their core homelands and conquer their way to the Arctic Ocean and the northern end of the Ural Mountains. But, by Ivan's time, even those vast lands were being trapped out. The increase in the wealth of Western Europe brought about by the conquest of the Americas led to new luxurious fashions in which fur played a major role. Even squirrel pelts were valuable. To put it another way, the value of fur was rising just as Muscovy was running out of it. The Western European eagerness to buy pelts created a great opportunity for the Russians to improve their access to Western markets if they could gain access to new fur-trapping lands. To solve his strategic and economic dilemmas, Ivan chose the path that many leaders both before and after him have chosen in times of stress and almost always with bad results: he declared war on everybody.

Ivan's first two campaigns were quick and successful. In the summer of 1552, he led his army down the Volga and laid siege to Kazan. Six weeks later his troops sacked the city and massacred most of the population thus gaining control of the western side of the Urals. Four years after that he continued down to the Volga and did the same to the Khanate of Astrakhan, gaining control of the entire length of the Volga River. With these two conquests, Ivan opened the way to the Caspian Sea and trade with Persia and the Middle East, increased access to untapped trapping country, and acquired rich agricultural lands in which to settle surplus peasants. Ivan now turned his attention to the Crimean Khanate, which blocked the way to the Black Sea. This led him into conflict with the Ottoman Turks, allies of the Krim Tatars, who sent an army across the sea and stopped his move in that direction. Undeterred, Ivan turned west and invaded Livonia (modern Estonia and Latvia), a principality that Poland and Sweden were in the process of dividing between themselves. The war was a disaster. Poland and Sweden had better armies, better weapons, and more wealth than Muscovy. Ivan's determination and increasing mania led him drag out his inevitable defeat for twenty-five years. At one point, even Denmark was at war with Muscovy. During the darkest days near the end of the war, Ivan found out that he was also at war with another neighbor, the Khanate of Sibir which dominated the land between the Urals and the Yenessi River.

Early in his reign and prior to the start of this seemingly perpetual war with the world, Ivan made the decision that would lead to Russia becoming the largest country on earth. In 1558, soon after conquering Kazan, he gave the Stroganov brothers, Grigori and Yakov, exclusive rights to exploit a territory the size of Maryland on the Kama and Chusovaya Rivers along the central Urals. In the space of less than fifty years, the Stroganov family had combined superb business acumen with uncanny political instincts to become the richest family in Russia and favorites of the tsars. Ivan had given them their fiefdom on the simple terms that they create prosperous new province for him by opening mines, starting businesses and trapping, and that they bear the expenses of guarding the border. Things went so well for the first sixteen years that, in 1574, Ivan renewed the Stroganov privileges and gave the family permission to look for additional opportunities beyond the Urals. Over the years, a small number of individuals had crossed the mountains and brought back enough fur that canny businessmen like the Stroganovs had a good idea of the opportunities to be found there. They just needed to find the right opening to exploit those opportunities.

In the fall of 1581, a large band of Volga River pirates, fleeing the law, followed the Kama River into the Stroganov lands. Rather than see a threat in the unruly force that had entered their lands, the sons of Grigori and Yakov saw the opening they had been looking for. They paid the leader of the pirates, Vasily Timofeyevich, called Yermak ("the millstone"), to take his men across the mountains to break the power of the khan of Sibir and open the east to the family's agents. On September 1, Yermak departed for Sibir at the head of a force of 540 Cossacks, three hundred Polish and Swedish POWs, two priests, and a runaway monk who had signed on as the cook. Hearing of the invasion, Ivan was enraged. By now, he was old, in ill health, and possibly insane. His war in the west had finally ended in defeat. The state was almost bankrupt. In a fit of rage, he had killed his eldest son and heir. And, to make things worse, the Stroganovs had stripped the border defenses to outfit Yermak. Siberian tribes were crossing other parts of the Urals unopposed and raiding eastern Muscovy. He sent messengers to accuse the Stroganovs of treason and to order the expedition back, but by the time the messengers reached him, Yermak had already conquered Isker, the capital of Sibir, and made a fugitive of the ruler Khan Kuchum. As the Stroganov cousins were making peace with their god and waiting for the executioner to come knocking at their door, Ivan Koltso, one of Yermak's crew, arrived at the Kremlin with an enormous load of furs and the message that the expedition had destroyed the khanate and annexed its lands in the name of the tsar. Ivan sent pardons, gifts, and honors to Yermak and his band. Most of the band and the first group of reinforcements died over the next few years trying to hang on to their conquest—Yermak drowned trying to swim in the gold chain mail Ivan had sent him—but, in the end, they were successful in adding a rich new province to the empire.

Yermak's expedition removed an enemy on one side of the kingdom and brought in unexpected new revenues. Over the next few decades, trappers, traders, and various malcontents, generically called *promyshleniki*, pushed the borders of Russia all the way to the Pacific Ocean and the frontiers of China. More than anything it was fur that drove that advance. For the eighty years following the conquest of Sibir, revenue flowing to treasury from the fur trade completely made up for the exhaustion of the old trapping grounds and steadily increased beyond it. But the methods of the *promyshleniki* were not sustainable. Their advance was so fast precisely because the trappers were exhausting one river basin after another in a frantic rush to stay profitable. The fur trade income peaked at about the time Witsen visited Moscow. Both the Siberian Prikaz and the settlers in Siberia were actively searching for new sources of wealth. One of the new items added to the state monopoly list was mammoth ivory.

Witsen discovered mammoth ivory during one of his clandestine visits to the market in 1665. At the time of his trip to Moscow, the Netherlands controlled all of the trade coming out of Ceylon (today's Sri Lanka), including its ivory trade. Witsen had seen and handled complete elephant tusks. This put him in an excellent position to evaluate Siberian ivory. What he found in the market was, according to him, "a little coarser and redder than fresh ivory" from Asia but he had no doubt that it was real elephant ivory. Of its origin, he writes: "Rivers in a certain Muscovite region, rushing down from the mountains often uncover on their banks heavy teeth, which are judged to be Elephants that were washed there at the time of Deluge and covered with earth: they are called by the Russians Mammotekoos. *Mammot* is Russian for a large dangerous animal and koos means bone." Witsen did not believe that the elephants from whom the ivory came had ever been native to Siberia; he agreed with the opinion that they must have drowned and had their bodies carried there by the biblical flood. To bolster the Deluge theory, he points out that Spanish travelers claimed to have discovered elephant bones and ivory in Mexico another land where elephants did not exist. Witsen wrote about *mammotekoos* in all four of the works he published in his lifetime.

The first place Witsen wrote about the ivory was in, of all things, a book on shipbuilding. Following the death of his father in 1669, Witsen discovered a great number of drawings and notes on naval architecture among the elder Witsen's papers. This inspired him to write a book on the subject. The book, *Aeloude and Hedendaegsche Scheepsbouw en Bestier* (Ancient and Modern Shipbuilding and Governance), was large and lavishly illustrated. The production was so difficult that two different publishers combined their resources to print it. This made it very expensive (twelve guilders or over a thousand dollars today). In spite of that, it was in very high demand. There was nothing else like it. The rich and powerful, even monarchs, bought copies and studied it, despite the fact that it was written in Dutch, hardly a universal language. The first part of the book is an overview of historical ships from Antiquity to Witsen's own time. This is where he mentions the mammoth. For the first two and a half pages of his book, he writes about sailing in the Bible. When he mentions Noah and the Deluge, he digresses to give some proofs of the Flood. First are reports of ships found buried or far from the ocean. Next, he mentions elephant remains in Siberia and Mexico. The only other proof he offers is the same seashells on mountains that had puzzled writers for the previous two thousand years. In 1690, Witsen released a second, expanded version of the book. In this edition, he added a description of a specific mammoth find near Kiev made five years earlier. The print run of this book was probably much smaller. He biographers weren't even aware of it until the early twentieth century.

In reading his book, many of the elite of Europe were first exposed to the idea of elephant bones in Russia and to the word "mammoth." How many took note of that fact is unknown. Probably not very many. Bones and seashells were not the reason they were reading the book; shipbuilding was. One person who did take note, however, was Leibniz, a correspondent of Witsen's. Leibniz was aware of the book months before it came out. When he was finally able to examine a copy four years later, he took extensive notes on the contents. The first time Leibniz mentions the mammoth passage is in the manuscript of his *Protogaea*, written in 1692 or 1693. By then, Witsen's third book mentioning the mammoth had come out, but the wording Leibniz uses exactly matches that in the two shipbuilding books. A second writer who learned about the mammoth from these books was Wilhelm Ernst Tentzel. Tentzel wrote a monthly journal on recent scientific advances, history, and other subjects he thought interesting. In his February 1690 issue, he mentions Witsen's book and the buried ships that appear in the same paragraph as the mammoth. His interest at the time was, as Witsen intended, as evidence for the Flood. Five years later he would be involved in his own controversy over buried bones and use

Witsen's description of the *mammotekoos* and bones in Mexico to bolster his case. Witsen, Leibniz, and Tentzel were regular correspondents with each other during that decade. Finally, someone with more influence than the three of them combined read the shipbuilding book and took notice. He had good reason to; he was Peter I, tsar of all the Russias, and Siberia was his personal property.

Peter the Great was born the year after the first edition of Witsen's book went on sale. While no monarch has what could be called a normal childhood, Peter's was exceptionally violent and tempestuous, even by the standards of the time. Peter's father, Aleksei, to whom Witsen had presented his credentials seven years earlier, was only the second of the Romanov dynasty. The position of the family was by no means secure and they had to constantly navigate their way among the old, powerful families of Muscovy. Aleksei married twice, first to Maria Miloslavskaya and second to Natalia Narishkina. Between them, he fathered sixteen children. At the time of his death, three males survived. Fedor and Ivan, the sons of Maria, were both in dangerously bad health. Peter, the only son born to Natalia, was strong as an ox, but only four years old and third in line. Aleksei's eldest son was sworn in as Fedor III, and died six years later leaving no heir.

This created a succession crisis. The Miloslavskis naturally supported Ivan while the Naryshkins supported Peter. Though the obvious choice should have been Ivan, the older of the two, his health was so bad that no one expected him to live very long. Ivan was severely epileptic and nearly blind, and may have suffered from a variety of other problems (diagnosing the physical and mental health of historical figures is more of a parlor game than a science among historians). Patriarch Joachim, head of the Russian Orthodox Church, sought to thread the needle by calling an assembly of the leading citizens of Moscow. The crowd chose Peter by acclamation. This type of royal election was common in Europe east of Vienna, but many cried foul because no time had been allowed for anyone but Muscovites to participate. Among the chief complainers were the *streltsy* (musketeers) a hereditary military caste created by Ivan IV. At the time, the *streltsy* were unhappy with the powers that be over a number of issues including late pay and abusive officers. The Miloslavskis fanned their discontent and played up their fears that Peter was a mere tool of corrupt forces around the court and that even darker plots were being hatched by the Naryshkins against the two princes.

Two weeks later, the *streltsy* rose up in bloody rebellion, mistakenly thinking they were protecting the monarchy. Two royal advisers were murdered in front of Peter and Ivan while Peter's mother tried to protect them. Peter would have nightmares about this for the rest of his life. Two of his Narishkin uncles were murdered. Another dozen men on a carefully prepared hit list were killed along with guards, family members, and a few people who were in the wrong place at the wrong time. In the chaos, Ivan's older sister, Sophia, emerged as the spokesperson for the terrified boys. At one point, she stood before a courtyard of angry *streltsy* and won them over by promising them their back pay. Four days later, she was one of those who oversaw a new royal election. The new election was dominated by the *streltsy*. At the last minute, they wavered about actually deposing Peter, a prince of royal blood, and elected Peter and Ivan co-tsars. Patriarch Joachim provided the appropriate historical justifications for such an unusual arrangement. Sophia would rule as de facto regent for her brothers.

Sophia was a remarkable woman. At a time when upper-class Russian women lived in seclusion and were rarely seen, she was able to address a courtyard full of angry soldiers and hold her own in negotiations with the most powerful men in the realm. Less than a week after her brothers' coronation, she presided over a debate between factions in the church. With the Narishkins in hiding and her sisters too traditionalist to appear in public, she was the only adult Miloslavski of royal blood available to look after the tsars. Historians have not been kind to Sophia. For most of the last three hundred years she has been stereotyped as a pushy, scheming, unattractive, and perhaps lusty woman who was finally put in her "place" by a strong male (Peter). The same terms have been used for most strong women of the modern era not named Elizabeth, Catherine, or Victoria. But Sophia was not just a remarkable footnote to history, she is important to our story.

During her regency, Sophia preferred to keep Peter and the surviving members of his family out of sight as much as possible. He lived in the village of Preobrazhenskoe with his mother and his tutor, Nikita Zotov. Zotov was an easy-going drunkard who let his charge do as he pleased.

Peter's intellectual strength lay in the technical arts. What pleased him was visiting with the inhabitants of the foreign quarter and learning about the outer world. They were happy to have him. The Scots and Germans taught him practical trades and modern military science, while the Swiss mercenary Fraz LeFort taught him about drinking and women. In 1688, Peter discovered a rotting English sloop on the lake at Preobrazhenskoe. A Dutch mercenary, Karsten Brand, helped him revamp it and taught him the basics of sailing. This was the beginning of Peter's lifelong enthusiasm for ships and naval power. Other Dutch residents introduced him to the basics of shipbuilding . . . and they had a very good book on the subject.

While Peter continued his eclectic education, interrupted by periodic journeys into the city for ceremonial occasions, Sophia tended to affairs of state. One of the crises she was called on to deal with was a little war in Siberia. Traders, trappers, missionaries, and bureaucrats reached the far end of Siberia long before sufficient numbers of farmers to feed them did. They traveled mostly on the rivers, by boat in the summer and sledge in the winter, building almost no roads on the boggy countryside. Even today, roads are rare in most of Siberia. Bringing large loads of bulky goods, specifically enough grain to feed a small settlement, was a difficult and expensive proposition. It could take three or four years for a shipment of grain to reach a remote place like Yakutsk, and by then, the majority of the load would be inedible. It was never a surprise to hear that a remote settlement had been devastated by starvation over the winter. Because of this, the *promyshleniki* were relieved and excited when they began to hear rumors of the Amur, a valley in the south filled with grain, cattle, and silver. Beginning in 1643, a number of Cossack groups attempted to conquer the valley, which the Chinese considered within their sphere of control. The resulting struggles were horrendous affairs involving kidnapping, plunder, and, it is reputed, cannibalism. The local Chinese authorities chased the Russians out of the lower and middle valley multiple times only to have them return as soon as the authorities were gone. In the 1660s, a group of Cossacks was able to establish a sort of free republic at Albazin, on the northern bend of the Amur.

The Kangxi emperor was unable to deal with their provocations because he had much more dangerous problems caused by rebellions in the south.

But the problems in the north were never far from his mind; the Amur valley was the homeland of the Qing dynasty. In the early 1680s, with China proper pacified, he turned his attention to the invasion from these Russian barbarians. In 1684, the emperor sent a large and well-supplied army to the lower Amur. The army methodically moved west, driving the Russians before them. The Russians attempted to make a stand at Albazin, but were soon defeated. The Chinese allowed the survivors to retreat and razed their fort. At this point they were poised to do some real damage to Russian interests in Siberia. Instead of exploiting their momentum to further advantage, the Chinese moved down river to their base of operations and waited to see what the Russians would do. The emperor sent messages to Moscow requesting that envoys be sent to negotiate a settlement that would be satisfactory to both sides.

Sophia understood that trade with China was far more important than the interests of a handful of out-of-control border ruffians. Since the beginning of the century, the tsars had been trying to open relations with China, but every attempt at making official contact had failed due to cultural misunderstandings. Sophia jumped at the chance to do what none of the men in her family had managed. After several delays, during which the promyshleniki rebuilt Fort Albazin, delegations from the two empires met at the Russian outpost of Nerchinsk on a tributary of the Amur almost three hundred miles west of Albazin. Sophia sent a large delegation that was met by a Chinese delegation ten times its size. Negotiations were carried out for the Russians by a Polish cavalry officer and for the Chinese by a French Jesuit. They negotiated in Latin and reached an agreement on August 27, 1689. According to the final settlement, Russia retreated from the valley and China agreed to allow regular trade through Nerchinsk.

Sophia did not get to celebrate the Treaty of Nerchinsk. At the same time that the negotiations were wrapping up in the east, Sophia's regency was coming to an abrupt and unanticipated end in Moscow. Sophia's position had been dramatically weakened by two disastrous campaigns in the Crimea and by her half brother Peter turning seventeen in June. Amid rumors that Sophia was planning to murder Peter and rule in her own name, supporters of the two Romanovs engaged in a month of dramatic maneuvers that resulted in Peter taking control and Sophia being packed off to a convent. Peter's half brother Ivan stayed on as co-tsar until his natural death seven years later from an unspecified illness. When word of the treaty reached the court, Peter and his advisers were thrilled at the opportunity and began planning a trade mission to Beijing.

Besides being acquainted with Witsen's pivotal book on shipbuilding from his childhood, Peter, possibly through Winius, became aware of Witsen's long study of Russia's borderlands. At some point in the late 1670s, while collecting information about northern Asia, or Tartary as it was usually called, Witsen came up with the idea of creating a new map of that part of the world. Europe knew virtually nothing about northeastern Eurasia. Though maps of the world and Asia had been published since the first part of the previous century, anything east of the Ob and Irtysh Rivers and north of the Great Wall of China was pure speculation. A vague coastline had been created based on the writings of Pliny and the land was filled with place names drawn from the Bible, classical geographers, and the fables of Marco Polo. Witsen's collected materials included enough sketch maps and travel itineraries, along with intelligence from his cousin, that he became confident that he had enough information about western and central Siberia to risk going public with a new map. Beginning in 1682, he printed a small number of maps and sent them to friends and prominent intellectuals around Europe, regularly making changes before more broadly releasing the map in 1687. The map was a huge success. Prominent map-makers used his data to update their own maps. He was not happy about what he saw as plagiarism, but he decided to let it pass for the greater good of spreading knowledge.

At least one copy of the map made it to the hands of young Peter. Among the many versions of the map is one dedicated to Peter. The design is identical to the 1687 version. The young tsar was delighted by it. Witsen, by then, had figured out that the future of Russia would belong to Peter. This was by no means guaranteed at the time, as at this point, Sophia was still the regent of Russia. Peter was only fifteen and the junior tsar next to his elder half brother. Despite his many medical problems, Ivan had married in 1684. Though childless at the time, he would father five children before his death. Sophia was still regent, and many believed she was planning to declare herself sole ruler of Russia.

The rumor happened to be correct; his cousin Winius was ordered to have a hundred portraits of Sophia printed; these were essentially coronation portraits. He sent the request to Witsen in 1687 to have them printed in the Netherlands. Winius must have sent other intelligence that led Witsen to believe Peter (or rather, his court faction) would prevail in the coming showdown with Sophia. Winius wasn't his only source of information. Twenty years had passed since his trip to Moscow and Witsen had cultivated an extensive network of informants. One of these was probably Prince Yakov Dolgoruky. Dolgoruky was a trusted member of the court. Later he would be a close confidant of Peter's, and, when he spent a month in Amsterdam on diplomatic mission in 1688, he and Witsen would spend many hours together speaking honestly and confidentially. Whatever the source of his faith in Peter, Witsen had a prominent dedication to Peter placed on the map's cartouche.

Witsen's five-year delay in releasing the map was due to more than perfectionism. There was a genuine risk in revealing the depth of his knowledge of Russia's eastern territories. Like the Spanish and Portuguese shipping routes of the previous century, such knowledge could be considered state secrets. Witsen's Russian sources could have been in real danger if their identities had become known. A twentieth-century study of the place names on the map showed that most were derived from Russian forms of the names, meaning he received most of his information from Russians and not from Western travelers. In the eighteenth century, Gerhard Friedrich Müller went through the Russian archives and decided that Winius was the source of most of Witsen's geographical information. Luckily for both Witsen and Winius, in his letter thanking Witsen for the map, Peter encouraged him to continue his researches. Witsen did just that. In hindsight, it makes perfect sense. The Russians did not have an accurate map of the east. They had sketch maps of small areas, and travel itineraries of the major rivers, but no one had assembled all of this information into a large modern map. This introduction led to a friendship that would benefit both Witsen and Peter for the next thirty years.

In 1690, Witsen received a letter in the name of the two tsars asking for advice on how to improve trade with China and Persia. Once again, cousin Andreas may have had a hand in this. With the shift of power,

Winius had become the head of the foreign ministry. By this time, Witsen had a title to accompany his intellectual prestige. He was one of the mayors of Amsterdam (there were four). He would hold the office thirteen times. His advice was only moderately useful. He extolled the virtues of free trade (such as opening Archangel to Amsterdam merchants) and urged a new war against the Chinese to attempt to retake the Amur delta. It was a major concession by the Russians to admit that a foreigner had more knowledge of the subject than they did. Peter's faction knew what a valuable resource they had in Witsen. During the coming years, they would turn to him for advice and for aid in trade, in hiring experts, and in acquiring arms for Peter's wars.

His new relationship with the highest circles of Russian power encouraged him to do something his friends had been requesting for years. He began organizing his twenty-five years of research into a book. *Noord en Oost Tartarye* (Northern and Eastern Tartary) came out in 1692, though, as with his map, "came out" is a bit strong of a description. Witsen had the book printed and bound, but kept all the copies and presented them to friends and respected others. In some cases, he gave the copies on the explicit condition that the recipients never let the copy leave their homes. Witsen's caution was based on his previous experience with plagiarism and intensified by his physical condition. He was over fifty, in increasingly ill health, and almost certainly clinically depressed. He wanted to be generous with his knowledge but feared others stealing his legacy. None of this stopped him from using the map to his political, financial, and intellectual advantage. He presented several copies of his book to influential Russians, including the two tsars to whom it was dedicated. Marion Peters, his biographer, suggests some other likely recipients: Feodor Golovin, who negotiated the Treaty of Nerchinsk; Matvei Gagarin, the future governor of Siberia, and cousin Winius. The book is divided into two parts. The first is an historical and political narrative of Siberia and the surrounding lands. The second part deals with the geography and resources of northeast Eurasia. Like Olaus Magnus had done in the previous century, Witsen links his description to his earlier map.

Witsen's book on the Tartary provides a great deal more information about the mammoth, which served to broaden the mystery of the animal without offering any solutions. Witsen begins by repeating what he said in his two shipbuilding books: spring floods uncover ivory on the banks of rivers, the ivory resembles elephant ivory, and the natives call it *Mammoutekoos*. A significant difference between this description and the version in his two shipbuilding books is that, although he still credits the Deluge with depositing the ivory in the north, he no longer draws a connection with the supposed elephants' bones found in Mexico. Additional information from Russia might have made cast some doubt on his previous conclusions. As he continues, he describes a second type of Siberian ivory: "At [the market] they also peddled the horns and teeth of the Behemoth, called by the Russians *Mammout* or *Mammona* . . . they say [*the mammona*] is dark brown and gives off a great stench: it is rarely seen and if seen betides many disasters: it has a tail like a horse's: short feet and other trifles that they believe." He tells his readers some of the products that the locals make from this ivory: "The Samoyeds make their arrowheads, and the Muscovites make all sorts of things which elsewhere are made from real ivory." He goes on to say that mammoth bones and ivory are mined from the ground and that the *Mammona* tooth is only nine inches long. By this point in his life he had been to India and seen live elephants, something that would have further influenced his theorizing on the mammoth. He has too much information at this point and is struggling to bring it all into a coherent whole. Rather than putting forth a solution that he has no faith in, he takes the honest path of giving his readers all of his data, however confused, and letting them draw their own conclusion.

Witsen mentions an alternative to the Deluge theory that, he says, some people think explains the presence of elephants in Siberia. He tells us: "Some Muscovites of that location [Siberia] are of opinion that by long ago the globe or world twisted, and that it is now cold where it had been hot, so that elephants, finding themselves in cold regions, ceased to exist." This idea of twisting the world to change the climate was something new when Witsen mentioned it. In 1681, Thomas Burnet published the first volume of his *Sacred History of the Earth*. In it, he attempted to explain the geology of Genesis and the world as we now see it through mostly naturalistic processes. In the original creation, Burnet told his readers, the earth was almost perfectly smooth and the pole was perpendicular to the ecliptic. There were no seasons, the climate was temperate in all latitudes all year around, and mankind lived in an eternal springtime. The Deluge was caused by the cracking open of the earth's surface allowing a layer of water between the crust and the core to surge to the surface in the form of "the fountains of the deep." The earth we see today is the wreckage of the Edenic past, covered with ugly heaps of stone and with the axis tipped to make great parts of the earth, in the north and south, barely habitable. Burnet's book was sixteen years in the future when Witsen visited Moscow in 1665. It's unlikely that the merchants he talked to at that date had independently arrived at the idea of a sudden shift in the earth's axial tilt changing climates. This explanation must have come from one of Witsen's informants at a much later date. The idea of combining mammoths with Burnet's theory would have come from a very westernized Russian, from one of the Westerners living in the merchants' quarter, or a Westerner employed by the government, such as Winius.

By 1692, Witsen's books had exposed a rarefied strata of European society to the word "mammoth" and to the idea of elephant-like ivory coming out of Siberia. However, in all three books, his mention of the mammoth was buried deep within discussions of other things. We know from his letters that some people noticed the mammoth and wanted to know more about it, but we can literally count those people on one person's fingers. Witsen's greatest contribution to understanding the mammoth wasn't in his own research; it was in using his intellectual prestige and connections to promote the next generation of thinkers.

Ever since word of the Treaty of Nerchinsk had arrived in Moscow, Peter's advisers had been planning a major trade and diplomatic mission to Beijing to test the terms of the treaty. Russia had a severe shortage of literate agents who were sensitive enough to make their way through foreign cultures without causing the same offenses that had brought failure on every previous mission to China. They turned to Witsen to recommend a leader for the mission. Witsen recommended Evert Ysbrants Ides, a German-Dutch merchant from Holstein. Ides was an excellent choice and had the advantage of already being known in Moscow. Ides was born in 1657, which put him about halfway between Witsen and Peter in age. He came from a family of small merchants—his grandfather was listed in the tax rolls as a Höker, a word that roughly translates as "grocer." By the time he was twenty, Ides was trading with Russia, sending ships from Hamburg and Amsterdam to Archangel. Although his firm traded as far away as Italy, he had enough faith in Russia as the future of his fortune that he purchased a home in Moscow and moved there in 1690. He became well known in the foreign quarter and even entertained the tsar himself in his home. Then he went bankrupt. Wars in Germany and the loss of ship with its paid-for cargo (and crew) destroyed his investments. Knowing of the terms of the Treaty of Nerchinsk, he petitioned the tsars for permission to lead a caravan to China and for a loan to buy some trade goods. A surprised Ides found himself drafted to be Peter's diplomat to the Chinese court and well compensated for the job above and beyond what he would have made on a simple trading expedition.

On March 14, 1692, Ides left Moscow at the head of a ninety-man caravan with instructions to exchange official ratifications of the treaty, determine the best items for trade, feel out official attitudes toward the treaty, and request that a Chinese envoy be sent to Moscow. Peter charged Ides with making careful observations of people and resources along the way and provided him with a secretary and an artist to aid in that task. Witsen sent Ides a version of his map both to guide him and so Ides could bring back additions and corrections for the next version of the map. The most direct route from Moscow to China is the one that the Trans-Siberian Railway follows today, around the southern end of the Ural Mountains, across the steppe lands at the center of Eurasia, across Lake Baikal, and on to the Amur. Unfortunately, the steppe lands were controlled by Kirghiz nomads and considered unsafe at the time. Ides's caravan had to take a much more roundabout path through the north to reach Lake Baikal. This route followed the same course that Yermak took across the central Urals to Tobolsk, the capital of Siberia, then down the Irtysh River to its junction with the Ob, up the Ob and its tributary the Ket, making a portage to the Yenisei basin, and finally following its tributaries upstream to Lake Baikal. By October 1692, the mission was at the way station of Makofskoi on the Ket portage. It was here that Ides heard an amazing story about the mammoth.

> I had a Person with me to *China*, who had annually went out in search of these Bones; he told me, as a certain truth, that

> he and his Companions found the Head of one of these Animals, which was discovered by the fall of such a frozen piece of Earth. As soon as he opened it, he found the greatest part of the Flesh rotten, but it was not without difficulty that they broke out his Teeth, which were placed before his Mouth, as those of the Elephants are; they also took some Bones out of his head, and afterwards came to his Fore-foot, which they cut off, and, carried part of it to the City of *Trugan* [Turukhansk], the Circumference of it being as large as that of the wast of an ordinary Man. The Bones of the Head appeared somewhat red, as tho' they were tinctured with Blood.

Ides goes on to tell of the different theories he had heard considering the nature of the animal. The "Heathens of Jakuti, Tungusi, and Ostiacki" believe that it is a living animal that burrows through the ground. As evidence, they point to places where the ground has been heaved up or sunk down by the animals on the march. Mammoths cannot breathe surface air. If they should accidentally come to surface, such as by tunneling out of a riverbank, they immediately die. Naturally, Ides gives more credence to the opinions of Christian Russians. From them, he was told the same story that Witsen recorded. Mammoths are a type of elephant that lived in Siberia before the Deluge when that part of the world was warmer. He concedes that this "is no very unreasonable conjecture," though he is more inclined to think that they were not local but, rather, elephants from India whose bodies had been carried north by the Deluge and left there when the waters receded.

Ides report has become quite famous because it is the first description of a specific frozen mammoth. When the Chinese chronicles mentioned the *fyn-shu*—if indeed that meant mammoth—they mentioned that they were found buried only as a general characteristic. They never described a specific discovery. When Witsen reported that the mammoth is "dark brown and gives off a great stench," his sources were probably describing a rotting carcass—later discoverers would complain about the horrible smell—but, again, this is a general characteristic, not the description of a specific mammoth. Unfortunately, the mission artist, Johann Georg

Weltsel, died of a fever a few days before the mission arrived in Makofskoi. If he had attempted to make some sketches based on the ivory hunter's description they would have been the first done since the mammoth went extinct 10,000 years earlier.

As a diplomatic effort, his three-year mission was a failure. The Chinese rejected the tsar's gifts. Peter didn't blame Ides for the failure, however, and was generally pleased with his work. After paying back the loans, Ides made a very nice profit and Peter continued to entrust him with important tasks. For a time, he managed a large portion of the Russian arms industry. Immediately after the voyage he was made the official representative of the Russian state for certain types of printing, including maps, in the Netherlands. Both of these enterprises would keep him in close contact with Witsen. Peter had given Ides permission to publish his journal outside Russia, and Witsen was eager to have it, but Ides was busy with his new duties and couldn't be bothered with the work of preparing it for publication. Several years would pass before Europe would hear his account of the mammoth. Meanwhile, Witsen was nurturing other travelers.

In the early 1690s, Heinrich Wilhelm Ludolf was the youngest of a family of well-known German intellectuals. His uncle, Hiob Ludolf, who raised him, was a diplomat and orientalist best known for his studies of Ethiopia and the Amharic language. Hiob took his nephew on his diplomatic missions to London where he introduced him to influential Englishmen and to other diplomats. During one of these trips, Heinrich secured a position as secretary to the Danish ambassador and soon became the personal secretary to Prince George of Denmark. He seemed destined for a brilliant diplomatic career. Then his life took an abrupt left turn. After five years with Prince George, Ludolf suffered some sort of mental breakdown. One of his friends wrote that he lost his reason after reading books by Rosicrucians. Whether or not the Rosicrucians were to blame, his unstable condition and increasing mystical obsessions made him an unsuitable secretary for a royal. Prince George had become quite fond of his secretary and made sure Ludolf would not be abandoned to the fates. He put him under the care of a highly respected doctor, provided him with a generous pension, and used his influence to smooth the way for Ludolf to follow his interests. In a way, Ludolf ended up living the dream of every underpaid academic.

As a young man, while living with his uncle, Ludolf had picked up an interest in and talent for languages. When he recovered his health, he decided to travel to Russia to study the language. To him, this was as much a religious enterprise as it was an intellectual one. Ludolf's pietistic beliefs placed a high value on missionary work. Successful missionaries should speak the local language and very few Westerners spoke Russian. Ludolf saw a need and set out to fill it. He left Denmark in the summer of 1692 and arrived in Russia at the beginning of the new year. He spent eighteen months there, mostly in Moscow, and made good use of the time. He became friends with Patriarch Adrian and the tsar. Peter was unconverted but enjoyed listening to Ludolf's musical performances on the bass. The time spent with religious figures aided him in teasing out the differences and relationship between spoken Russian and the liturgical language Old Church Slavonic. Ludolf left Moscow in the summer of 1694 and arrived in Amsterdam that October, where he was met by Witsen. Witsen was highly impressed by Ludolf's work and urged him to publish it. When they heard about his project, his old friends in England were eager to see it and arranged for the Oxford Press to publish it. The result was *Grammatica Russica*, the first systematic study of spoken Russian published in any language, including Russian. Ludolf wrote it fairly quickly while staying with Witsen, but there was a delay in publishing because no printer in the British Isles owned a Cyrillic typeset. Witsen knew the publishing industry well and had previously acquired a Coptic typeset for Oxford. He had no trouble locating a Cyrillic one, which he bought and sent over with Ludolf and the manuscript. The book was ready for sale in May 1696.

The word "mammoth" appears near the end of the work. Following the main grammar and a list of common phrases useful to the traveler, Ludolf added an appendix of natural science terms. "Mammoth" appears in the section on minerals, because mammoth ivory was found buried in the earth.

> The *mammoutovoi kost* is a thing of great curiosity, which is dug out of the ground in Siberia. The vulgar tell wonderful stories about it; for they say that the bones be those of an

> animal which burrows in the ground, and in size surpasses all the creatures living on earth's surface. They administer them medicinally for the same purposes as they do that which is called the horn of the unicorn. . . . [T]he more skillful tell me that these *mammoutovoi kost* are elephants' teeth. So that it appears necessary that they were brought thither by the universal deluge, and in the lapse of time have been more and more covered with earth.

From this point on, the word and idea of the mammoth began to spread fairly quickly through the Republic of Letters. In 1697, Robert Hooke, the brilliant but argumentative curator of experiments for the British Royal Society, mentioned the word in a public lecture: "We have lately had several Accounts of Animal Substances of various kinds, that have been found buried in the superficial Parts of the Earth . . . , [such as] the Bones of the *Mammatovoykost*, or of strange Subterraneous Animals, as the *Siberians* fancy, which is commonly dug up in *Siberia*, which Mr. *Ludolphus* judges to be the Teeth and Bones of an elephant." By then, Witsen had grown tired of waiting for Ides to publish his journal of the mission to China so he encouraged the mission's secretary, Adam Brand, to publish his shorter journal. In 1698, Brand did just that in his native German. His version of the journal did not include the passage on the mammoth discovered near Turukhansk. However, as an appendix, Brand added Ludolf's natural history lexicon with its entry on the mammoth. Over the next year, English and French translations were published with help from Witsen.

In early 1696, just three months before Ludolph's *Grammatica* was published, Ivan V, Peter's brother and co-tsar died, leaving Peter the sole autocrat of one of the largest empires to have ever existed. Until then, Peter had allowed his circle of advisers, which included his mother, to represent him in most things while he gave his attention to pet projects, such as modernizing and building up his military. Not everyone was happy to have the foreigner-loving tsar as their uncontested ruler. Peter dealt harshly with complainers and plotters. A year later, when he felt secure on the throne, he shocked the court by announcing he would be leaving the country and visiting the west for over a year. Peter's absence was officially a secret; he did not believe his countrymen would respond well to his abandoning them to consort with even more foreigners. In March, the Grand Embassy left Moscow. It was officially headed by three of his most trusted companions. The 250 members included sons of nobles, guards, musicians, five dwarves, and sixty-seven "volunteers" who were to learn valuable technical skills. Among then was a six-foot-eight cadet named "Peter Mihailov," who absolutely was not the tsar.

After crossing northern Germany and conferring with great rulers and thinkers, the embassy reached the Netherlands in August. Peter ignored the government, went straight to the shipbuilding town of Zaandam, and enrolled as an apprentice carpenter. His attempt to learn the craft was a disaster. Every day, large crowds gathered to watch the Russian giant, who absolutely was not the tsar, attempt to work. After a week, he gave up and fled to Amsterdam where he was met by Witsen. Witsen found a house for Peter and arranged for him to apprentice on a frigate being constructed by Gerrit Pool at the Dutch East India Company's shipyard. Here he could learn in peace, because the shipyard was closed to the public. During the five months that Peter stayed in Amsterdam, Witsen scheduled his work so that he would spend one day each week with the tsar. He also accompanied him whenever he left the city.

Peter was fascinated by the cabinets of curiosities that he he saw when visiting upper-class Europeans. It might have been in this context that he and Witsen discussed mammoths. In a letter written to Gjisbert Cuper the following July, he reported that the tsar had himself seen mammoth bones along the banks of the Don River during a military campaign against the Turks at the mouth of the river. Thirty leagues below Olonets he and his entourage—a group that probably included Ides—saw that the river had uncovered a great mass of bones, including those of men and elephants. The tsar believed them to be the remains of part of Alexander the Great's army that had come north to fight the Scythians. Fedor Golowin, one of the official heads of the embassy, told Witsen that he had frequently seen mammoth bones during his time as governor of Siberia. Later, Peter would order the chief administrators in Siberia to pay attention to mammoth reports and offered a reward to anyone who could find a complete mammoth skeleton and bring it back for the cabinet that he was assembling.

In his letter to Cuper, Witsen recounts two other stories about mammoths that do not appear in *Noord en Oost Tartarye.* The first is the story told by Ides's traveling companion, as Witsen possessed Ides's manuscript by this time. Too busy to deal with it himself, Ides sent his journal, Witsen's map, and all of the notes concerning improvements to the map to him in one large, unedited package. In 1697 Witsen corresponded with Leibniz about Ides's observations on Siberian ethnography. With Witsen, the problem publishing the journal was that it was written in the Low German dialect spoken around Hamburg and not in proper, literary High German. After the turn of the century, he found the time to edit the journal, translate it into Dutch, and have it published in 1704. The book was very well received. So much so that an English edition appeared in 1705 and a German one in 1707. Ides's story of the frozen mammoth would become a standard part of mammoth lore. Yet Witsen had a low opinion of his second new source. In his letter to Cuper, he refers to the tale told by a Jesuit named Avrie as "ridiculous and untrue" and "a decorated lie." What offended him was that the good father seemed to be saying that there were still living elephants in northern Siberia.

How did the Jesuit Avrie come to this conclusion? In 1681, seventeen years before Witsen wrote to Cuper, a letter from Ferdinand Verbiest, the head Jesuit at the Kangxi emperor's court in Beijing, arrived in Europe and called on his Jesuit brothers to join him in China. The Vatican and the Order of Jesus were happy to fill his request but were unsure how best to do that. The sea voyage was so perilous that only one in three missionaries made it to China alive. What was needed was a safe land route to China. They knew it was physically possible; merchants and diplomats had journeyed safely there and back in the past. What they needed to know was whether it was politically possible. With this goal in mind, they decided to send a French member of the order, Father Philippe Avril (not Avrie, as Witsen called him), and four companions to reconnoiter the unknown middle of Eurasia. Western Europeans were still uncertain about the inland extent of the Chinese Empire, and his superiors greatly underestimated the distance involved. Avril's journey began in 1685 and took him through Rome, Cyprus, and Syria to Armenia where he met Louis Barnabe, another Jesuit with extensive knowledge of the Central Asian trade routes. Avril was undeterred when he learned the true length of the journey. He was happy for the opportunity to preach the Gospels to "the Barbarians" along the way. Barnabe joined him and the group continued on to the Caspian Sea, where they sailed to the mouth of the Volga and up to Astrakhan. Avril tried to join several caravans going east but was denied permission by the local authorities every time. After several months of delay, they gave him permission to go to Moscow and plead his case. After another full year of delay in Moscow, the government not only refused him permission to go any farther east, it ordered him out of the country. Avril's eviction had little to do with his mission. Sophia, who was nearing the end of her days as regent for her brothers, had recently sent ambassadors west to recruit allies for another campaign against the Turks and Krim Tatars. She felt her ambassadors had been badly treated by the French court, and Avril was the first Frenchman available to whom she could return the insult.

The mission wasn't a complete failure. Like Witsen before him, Avril used his year in Moscow to gather information about the east from merchants. And, like Wisten, hidden among the intelligences about Siberia was a description of the source of Siberian ivory.

> Besides furs of all sorts . . . they have discover'd a sort of Ivory, which is whiter and smoother than that which comes from the *Indies*.
>
> Not that they have any Elephants that furnish 'em with this Commodity (for the *Northern* Countries are too cold for those sort of Creatures that naturally love heat), but other Amphibious Animals, which they call by the Name of *Behemot*, which are usually found in the River *Lena*, or upon the Shores of the *Tartarian-Sea* [Arctic Ocean]. Several teeth of this Monster were shewn us at *Moskow*, which were ten Inches long, and two at the Diameter at the Root . . .
>
> But certainly nobody better understands the price of this Ivory than they who first brought it into request; considering how they venture their Lives in attacking the Creature that produces it, which is as big and as dangerous as a Crocodile.

Later, on his way out of the country, Avril met with I. A. Musin-Pushkin, the governor of Smolensk, who had served a stint in Siberia. Musin-Pushkin gave Avril some additional intelligence on his *Behemot.*

> There is, said he, beyond the Obi, a great River call'd *Kawoina*, into which another River empties it self, by the name of *Lena.* At the mouth of the first river that discharges itself into the Frozen Sea, stands a spacious Island very well peopl'd, and which is no less considerable for hunting the *Behemot*, an amphibious Animal, whose Teeth are in great esteem.

Several things stand out in Avril's account. His physical description of the ivory he saw and the amphibious animal that produces it is clearly that of a walrus. His transliteration of the word "mammoth" (or *mamout* or *mamant*) as *behemot* was something Witsen reported in *Noord en Oost* Tartarye. Walruses were then hunted in the Laptev Sea near the mouths of the Lena and Kaiwona (or Olonets). That he says the ivory was also hunted in the Lena is telling. The lower reaches of the Lena River are one of the richest grounds in Siberia for collecting mammoth ivory. Walruses live near salt water; they stay on the coast and don't move up rivers. Avril was doing more than simply applying the *behemoth* name to the walrus. What he understood was a mash up of two sources of ivory, walruses on the coast and buried mammoth remains on the riverbanks. This is very much like Witsen's two types of mammoth in his book. Avril's account was first published in French in 1691. Within two years it had been translated into English, with the Protestant translator issuing an obligatory denunciation of Avril's Catholicism before enthusiastically recommending the work.

In 1705, Witsen began circulating pages of an updated edition of *Noord en Oost Tartarye* with a considerably expanded section on the mammoth. In this version, he repeats Ides's story as well as almost everything Avril had to say on the topic while refraining from criticizing him. He must have reread the good father's account while preparing the manuscript and recognized the similarities between their Behemoths. Because he gathered so much information about the mammoth into once place, the second edition of *Noord en Oost Tartarye* could have been of great use to anyone interested in the mammoth question, but Witsen printed only a few copies for his friends. And still he collected information from the east. In 1709, one of his Russian contacts sent him a complete mammoth's jaw. He optimistically had a drawing made of it for an anticipated third edition of his book. He never completed that revision. He died in 1717. There is an apocryphal story that Peter was at his bedside at the end, but it's sadly not true. Peter was in the Netherlands at the time. He was deeply saddened that he couldn't have been there to say goodbye to his ally, mentor, and dear friend.

In the last years of the seventeenth century and first years of the eighteenth, the mammoth became a permanent part of the world known to European intellectuals. Every report of the mammoth during those years was in some way touched by Nicolaas Witsen. He deserves some sort of title for that. In those years, several theories were proposed to explain the nature of the animal and meaning of the word. It was an elephant. It was a walrus. It was an animal native to Siberia. It was a tropical animal carried north by the Deluge. Witsen left the Republic of Letters with more questions than answers. Before Witsen, people did not even know to ask questions about these mysterious beasts. Now that they knew of the idea of the mammoth, they wanted to know more. More information would come, and when it did come, it would come in a sudden rush as the result of a national tragedy.

CHAPTER 4

THE SWEDES

The Battle of Poltava (June 27, 1709) is a regular inhabitant of lists of great battles that changed the world. At this battle the armies of Peter the Great of Russia decisively defeated those of Charles XII of Sweden outside a small village in the free Cossack territory, which is now part of Ukraine. Charles, badly wounded, fled the battlefield and sought asylum in Moldavia, a tributary of the Turkish sultan. His small escort was the only part of his army to escape death or captivity at Russian hands. The bulk of his army attempted to follow him into Turkish territory, but, four days after the battle, they were surrounded by Russian cavalry and negotiated a surrender. Of the thirty thousand who began the campaign only five thousand survived to return home after peace was concluded twelve years later. The battle marked the end of Sweden as a great European power and the beginning of Russia's geopolitical ascendancy. Somewhat ironically, this national tragedy for Sweden and personal tragedy for the prisoners and their families turned out to be a bonanza for European science and the study of the mammoth.

One of the many, many injustices of war is the uneven treatment of prisoners. Officers are almost never treated as badly as enlisted men. The prisoners of Poltava were no exception to this rule. The common Swedish soldiers were marched off to labor in the frozen swamps of the Neva delta building Peter's new capital city. For them, captivity was brutal and most died before the final peace was signed and they could be freed. For educated officers, captivity was, in the words of one, "not awful." Officers were treated as exiles, not prisoners. This meant they were sent to a town and given full freedom of movement within the boundaries of their district. They took jobs and made significant contributions to the local economies. Peter took advantage of this sudden influx of so many Western-educated men. He gave them the freedom to practice their religion, build churches, and even to proselytize. The officers set up a school that was patronized by some of the most important families of Moscow. Many officers were put to work surveying the resources of the empire. Some married into the local population. When the war was over, Peter asked many to stay and continue working for him. The majority politely declined.

Most of the places of exile were towns near the Ural Mountains. The largest number of the Swedish officers ended up in Tobolsk, where eventually they made up a quarter of the population. The governor of Siberia, Prince Matvei Gargarin, was a man of intelligence and curiosity (and fantastic corruption). He sent his prisoners to explore places as far away as Kamchatka on the Pacific, the high Arctic, and deep into Central Asia, where some were captured a second time by soldiers of the Djungar Khanate. Gargarin was most interested in mapping his gigantic province and surveying resources (you can't steal what hasn't been found yet). Scientific knowledge came second. Everywhere the Swedes went, they heard about the mammoth and saw mammoth ivory. Some lower-level officers even made a living as ivory carvers and, thus, learned a great deal about the properties of mammoth ivory, such as the many colors it came in, the texture, and the characteristic twist of complete tusks. This new knowledge was not disseminated westward into Europe as it was collected. Instead, most of it was released in one sudden rush when the prisoners returned home following the Treaty of Nystad in 1721, many of them bringing mammoth ivory with them.

The one exception to this delay was Captain Johann Bernhard Müller who had been captured after the defeat at Poltava and sent to Tobolsk along with the other Swedish officers. In 1712, the tsar sent Müller down the Irtysh and Ob Rivers to study the Ostyaks (now known as Khanty) who inhabited the territory between Tobolsk and the Arctic Ocean. The metropolitan of Siberia, Philotheus, was involved in a major campaign to convert the Ostyaks at the time. When their paths crossed, Müller was able to compare travel notes with the metropolitan's assistant, Grigory Novitsky. Parts of their two reports are so similar, that it's easy to suspect Müller plagiarized Novitsky. Or perhaps the did collaborate, although that seems unlikely. Whatever the truth is, it was Müller's report of the region, including plentiful descriptions of mammoth ivory, that got Peter's attention and not Novitsky's. The tsar was so pleased with it that he allowed it to be published abroad even before Müller was allowed to leave the country. Perhaps it was his flattering dedication to Tsarina Catherine that did the trick. The report was published in German in Berlin in 1720, in English the following year, and in most of the other major European languages before the end of the decade. Novitsky's report went into the archives and wasn't rediscovered and published until 1884.

Müller's account of the mammoth has many aspects that make it stand apart from other accounts. He makes no mention of the Behemoth and doesn't speculate about the origin or meaning of the word, which he gives as "Mamant." He also only uses the word as the name of the ivory and gives no name to the animal that produced mamant: "There is a Curiosity in *Siberia*," he writes, "no where else to met with in any Part of the World, for ought I know. This is what the Inhabitants call *Mamant*, which is found in the Earth in several places, particularly in sandy Ground. It looks like Ivory both as to Colour and Grain." Müller reviews the various explanations for the ivory. To him, it is impossible that it came from real elephants. Siberia, in general, is too cold for elephants and the ivory is most plentiful in the coldest parts of Siberia. That it might be ivory of tropical elephants washed there by the biblical Flood "is so absurd, that it needs no further Refutation." Apparently, it did need refutation. Müller was almost alone in his generation in rejecting that idea out of hand. He says that he was originally attracted to the idea that it was *Ebur fossile*, or a trick of nature. After all, didn't salt "grow" in the earth that was just as good as real salt from the sea and didn't coal grow in English soil, too? (Novitsky used these same examples) What changed his mind about mammoth ivory, though, was hearing about ivory discovered with skulls and other bones that were still bloody. If the source of mamant was a real, living animal, what sort of beast was it? This is Müller's real contribution to mammoth studies.

> I have spoke to many Persons, who averred to me for the greatest Truth, that beyond the *Beresova*, they saw such Beasts in Caves of the high Mountains there, which are monstrous according to their Description, being four or five Ells high, and about three Fathoms long, of a greyish Coat, a long Head, and a very broad Forehead, on both sides of which just above the Eyes, they say, stand the foresaid Horns, which it can move, and lay cross-ways over each other. In walking it is said to be able to stretch it self to a great Length, and also to contract it self into a short Compass. Its Legs are, as to Bigness, like those of a Bear. Notwithstanding these Accounts, one does not know how far to rely on them, for that Nation do not trouble themselves about exact Enquiries of that Nature . . .

The real avalanche of new information didn't begin to arrive until two years after Müller's little book was published. In September 1721, peace was finally concluded between Russia and Sweden with the signing of the Treaty of Nystad. The Swedish prisoners were free to go home, but this would take some time. Travel in Russia was a very primitive affair, and Peter left the former prisoners to their own devices, providing them with very little help beyond traveling papers. It took months for them to find out that the war was over and even longer to leave the country, since they were expected to pay their own way. Some took as long as two years to get home. The expedition that had been captured by the Djungars didn't return until 1733.

At the December 14, 1722, meeting of the Royal Swedish Literary and Scientific Society held in Uppsala, Erik Benzelius, one of the society's founders, exhibited a drawing that had been sent to him by Baron Leonard Kagg, an officer just returned from Tobolsk. The drawing shows an animal with a cow-like body, long horns twisted around each other, a lion's tail, and large feet with long, curved claws. It is reclining in a manner similar to Scythian and other Inner Asian animal art. Kagg wrote that the drawing, labeled "Behemoth" and "Mehemot," was of a mammoth. The drawing came with the following description written on the back:

> The length of this animal, called Behemot, is 50 Russian ells [117 feet]; the height is not known, but a rib being 5 arshin long [or 11.7 feet], it may be estimated. The greatest diameter of the horn is half of an arshin [14 inches], the length slightly above four; the grinders like a square brick; the foreleg from the shoulder to the knee 1 ¾ arshin long [4 feet], and at the narrowest part a quarter in diameter. The hole in which the marrow lies is so big that a fist may be inserted, otherwise the legs bear no proportion to the body, being rather short. The heathens living by the River Obi state that they have seen them floating in this river as big as a "struus," i.e. a vessel which the Russians use. This animal lives in the earth, and dies as soon as it comes into the air.

This was not the society's first encounter with the mammoth. The minutes of the meeting indicate they were already familiar with "Capt. Müller's account of the Ostiaks" and descriptions by several other returning officers. Kagg's drawing should have added to their knowledge and answered their questions. Instead, it confused them and created new questions. It was generally accepted as a law of nature than no animal had both horns and claws. Based on reports that bones and ivory of the mammoth were usually found on riverbanks and near the Arctic Ocean, Olaus Rudbeck and Peter Martin thought that the animal might be a *siödiur*—a sea monster—and not a land animal. By now, they heard the animal called *mehemot, mamant, mammut*, and Behemoth. Which was correct? Could the name shed some light on the nature of the animal? Many theologians of the time were of the opinion that the biblical Behemoth was a hippo (they still are). Not knowing very much about hippos, some members wondered if the Siberian animal was a hippo. The meeting adjourned with a resolution to write to Baron Kagg for additional information.

Olaus Rudbeck was open to two possibilities about the mammoth. Looking at Kagg's drawing and the description on the back, he could see a possible connection in the *siödiur* legend. But, looking at other reports that described elephant-like features in the mammoth, he saw another possibility, one he inherited from his father. The elder Olaus Rudbeck arrived on the scene just as Sweden's power peaked. Many intellectuals felt the kingdom needed a suitably glorious past to match its glorious present. France could point to Charlemagne, England could point to Arthur, Italy could point to the Romans, and Scotland could point to the Picts who stopped the Roman Empire. Even the Greeks, languishing under Turkish domination, could claim to be the very fountain of Western civilization. Swedish history went back a few centuries and then fizzled out. No one was particularly impressed with pagan Vikings at the time. When Rudbeck arrived at Uppsala University, the primary effort to fill this lacuna was being made through linguistics. Like Gorp did with Dutch in the previous century, a circle around Georg Stiernheilm was trying to prove that Swedish was the original language of Eden and Noah and that all languages were derived from it. Also, as with Gorp, their proof consisted mainly of tortured etymological comparisons with Hebrew. Rudbeck happily accepted this theory and made a glorious addition to it. At some point in the 1690s, he took a good look at the landscape around Uppsala and was amazed to realize that it exactly matched Plato's description of Atlantis. Later, a nephew would be equally amazed to discover Troy in southern Sweden. This was the intellectual atmosphere in which the younger Rudbeck was raised.

The younger Rudbeck's linguistic theory was a tiny bit better grounded than Stiernheilm's. In 1695, he made a trip into Finland and Lapland to study natural resources and became acquainted with the languages there. About the same time he considered the mammoth problem, he reversed the cause and effect relationship linking the northern languages and Hebrew.

Rather than Hebrew being the descendant of Swedish, he proposed Hebrew as the source of Finnish and related languages. According to the canonical books of the Old Testament, after the division of Solomon's kingdom, the Assyrian Empire attacked and annexed the northern kingdom, carrying off the ten tribes that had inhabited it. From there, they vanish from history. In the Apocryphal book 2 Esdras, the ten captive tribes were allowed to leave and migrated eastward through the land of Arzareth, beyond the bounds of the known world. Renaissance mapmakers placed Arzareth in Ukraine or South Russia and scattered the tribes around northern Eurasia. Rudbeck learned from Captain Tabbert that Finnish, Lapp, and Estonian were related to other languages that extended across northern Europe deep into Siberia. He determined that these must be the ten lost tribes. From these diverse theories, Rudbeck concluded that the lost tribes used elephants as beasts of burden on their migrations into the most extreme north. Once there, being suited to a warmer climate, their herds of elephants eventually died, leaving only their bones.

When the society returned after the New Year, they once again took up the question of the mammoth. Martin reported that he had carefully examined the works of zoology, but could not find an animal like the one in Kagg's drawing though, in his opinion, it most resembled the Nile hippopotamus. The image of a hippo that he referred to is found in Hiob Ludolf's *Historica Aethiopica*. It is indeed a fierce and dangerous looking animal. Benzelius announced that Lieutenant Colonel Peter Schönström, Charles XII's secretary was bringing a mammoth tooth from Siberia and would bring it to the society when he arrived. Johan Malmström had a letter from the highly respected linguist Johan Gabriel Sparwenfeld copied into the minutes. He was of the opinion that "mammoth" could be derived from "Behemoth" using sound linguistic principles. When he was younger, Sparwenfeld spent four years in Moscow studying the Russian language. While there, he had acquired a piece of mammoth ivory. He reported that it was exactly the same as elephant ivory. This did not conflict with his Behemoth statement; Sparwenfeld was of the opinion that the word "Behemoth" referred to the elephant and not to the hippo.

In February, Benzelius journeyed to Stockholm to attend a session of parliament. While there he was able to meet with Baron Kagg in person and ask the society members questions. Unfortunately, most of the answers were "I don't know." Kagg explained that he wasn't the one who made the drawing; it had been given to him by another officer in Tobolsk. He couldn't be sure, but he thought it had been Captain Tabbert. The society would have to wait a bit longer to settle the mystery of the drawing. When the war ended, Captain Tabbert was deep in Siberia and by the beginning of 1723 he had only made it as far as Moscow on his return voyage. As something of a consolation prize, Benzelius was able to acquire another mammoth tooth. This one came from Gustaf Adolf Clodt von Jürgenburg, who had earlier sent the society a collection of Scythian antiquities gathered during his time in Russia.

Captain Tabbert was Philipp Johann Tabbert von Strahlenberg from Swedish Pomerania, a province in Germany that had been conquered during the Thirty Years War. In 1695, he and two of his brothers joined the Swedish army. During the war, the Tabbert brothers rose quickly through the ranks, driven by their education and intelligence. Philipp went from fortification engineer, to quartermaster, to captain of a signal regiment. In 1707, citing valor in combat, Charles ennobled all three brothers with the name Strahlenberg. It is by this name that we'll refer to him from here on. Phillipp Strahlenberg and his brother Peter were both at the Battle of Poltava. Peter was with the king's escort and escaped while Phillipp was captured and eventually ended up in Tobolsk. Like Witsen before him, he became fascinated with the country and set about learning everything he could. Tobolsk was an ideal place to do this. As the capital and main city of Siberia, it was the center of trade and the destination for bureaucratic reports. Phillipp Strahlenberg interviewed everyone he could. When Swedish officers were sent out of the district on various tasks, he had them collect data for him. By the time the war was over, Strahlenberg had collated a comparative lexicon of thirty-two Asian languages. He collected rubbings of ancient petroglyphs, samples of Inner Asian alphabets, and drawings of native costumes. He drew a map of Siberia and Central Asia that was more accurate than anything published in a generation. He was one of the first Westerners to describe shamanic rituals and the first to describe the psychoactive properties of magic mushrooms (*Amanita muscaria*). He wrote a short encyclopedia of the economy and curiosities of Siberia.

During the summer of 1720, a gloomy German doctor named Daniel Gottlieb Messerschmidt arrived in Tobolsk on a mission from the tsar to conduct a survey of medicinal plants and resources in Siberia. His written instructions also directed him to gather "all the curiosities to be found in the region of Siberia, including objects of antiquity, pagan idols, [and] large mammoth bones." Messerschmidt spent the winter in Tobolsk preparing for the expedition and gathering information from the locals. By the end of the winter, he had come to regard Strahlenberg as an indispensable intellectual companion and gained permission for him to accompany him on his travels as his assistant. During the year the two spent on the road, Strahlenberg and Messerschmidt actively investigated the mammoth question. They gathered mammoth bones in addition to the valuable ivory. No doubt, they discussed their ideas about the beast while they were on the road. It was Strahlenberg's bad luck that the one time he could have seen mammoth remains in situ, he and Messerschmidt were traveling separately and only the doctor was able to examine the site. In May, 1722, they arrived in Krasnoyarsk where Strahlenberg learned that the war was over and he was free to leave Russia. Messerschmidt entrusted him with the notes and samples that they had collected so far, asking him to deliver them to the university in Moscow. He also gave him two mammoth's teeth that he asked him to send to his friend and colleague Johann Philipp Breyne in Danzig. With that Strahlenberg turned west on the trip home that would take him more than a year to complete and Messerschmidt headed east and deeper into Siberia.

Strahlenberg arrived in Stockholm in June 1723 and was immediately invited to dinner with the society. Strahlenberg dazzled the members of the society with tales of "strange ice caves, about all sorts of unfamiliar fruits and trees, petrifications, reindeer, mountain goats, elk, and deer on the Yenisey, Tomski and several other rivers, and of Ostiaks on the Obi River, who said they 'come from a country-called Suomi-roll, which could be none other than Finland.'" If he talked about mammoths, it was not recorded in the minutes of the society. Perhaps mammoths are the petrifactions mentioned. Perhaps he talked about them along with the other animals along the great rivers.

Though the society's minutes failed to record anything, Strahlenberg's thoughts on the mammoth are not lost to us. He was given housing with his regiment and, once settled in, began organizing all the materials he had collected for publication. Many looked forward to his book. When John Bell, a Scotsman on a mission to China for the tsar, passed through Tobolsk in 1719 he wrote in his diary "Captain Tabar, a Swedish officer was at this time writing a history of Siberia. He was a gentleman capable for such a performance, and if it shall ever be published, it cannot fail of giving great satisfaction to the curious." But it would be seven years before the book finally saw the light of day as Strahlenberg met with repeated setbacks. His original plan had been to organize the book around his map, as had Magnus and Witsen. The first draft of the map that he drew in Tobolsk was stolen. The second draft was confiscated by Governor Gargarin. Strahlenberg was reunited with this map on his way out of Russia but had to sell it for traveling money. While organizing his book, several maps were printed in Europe that were based on his two earlier maps. Strahlenberg postponed his book while he gathered new information to make a better map.

Another part of the book was to be a translation of a Tatar chronicle that he and Peter Schönström had worked together on. In 1726, this translation was published anonymously in Leiden. Schönström assumed Strahlenberg had betrayed him and the two never spoke again. Strahlenberg returned to his collections to find other materials to make the book worthwhile. When *Das Nord-und Ostlische Theil von Europa und Asia* was published in 1730, it consisted of the map with some comments, his linguistic materials, a history of Russia and its leading families, and the little encyclopedia of trade. In the last part, we find a four-page entry for "Mamatowa Kost, which the *Germans* call *Mamot's* Bones or Teeth."

Strahlenberg uses the word "mammoth" to describe the whole animal, not just the ivory. He begins by repeating many of the points made by the others. The bones and teeth of mammoths are found in the spring when the floods wash them out of riverbanks. The teeth can be made into anything ivory can. He is sure that it is real ivory and not a product of the earth. With his interest in languages, it's no surprise that he goes into the etymology of the name. Strahlenberg is sure the name derives from Behemoth by way of the Arabic form *mehemot*. *Mehemot*, he explains, is sometimes used as an adjective to describe something large. Arab merchants arriving northern Asia and seeing the gigantic remains of an unknown animal called the bones and teeth *mehemot*. The Tatars naturally adopted the word thinking it was a proper name. In turn, the Russians arrived in Siberia, learned the name from the Tatars and corrupted it into mammoth. Having satisfied himself as to the origin of the name, he finally addresses the real question: what kind of animal was the mammoth? "But this is not so readily answered," he admits.

Strahlenberg rejects out of hand the idea that the bones could have come from an amphibious animal living in and near rivers. He admits that he, like so many others, once thought they were the bones of elephants washed into Siberia during the Deluge but, having learned more, no longer believes that. The size, proportions, and curve of the tusks are all wrong for elephants. In support of this he repeats a story told to him by an "ancient Painter, one Remessow." Several years earlier, Remessow and a group of companions discovered a skeleton between the towns of Tara and Tomsk in southern Siberia. The skeleton was "thirty-six Russian Ells long [88 feet], lying on one Side; and the Distance between the Ribs on one Side, and the other, was so great, that he, standing upright, on the Concavity of one Rib, could not quite reach the inner Surface of the opposite Rib with a pretty long Battle-Axe which he held in his Hand." For Strahlenberg, the only remaining possibility was that the mammoth was some kind of sea creature similar to a narwhal. He gives as his reasons the facts that most mammoth bones are found on river banks and that they become more plentiful the closer you get to the Arctic Ocean. He repeats stories of whales swimming up rivers during the flood and getting stranded. To account for mammoth bones found farther from the sea he says that perhaps before the Deluge the Arctic Ocean extended further inland. He offers this conclusion as the only possible solution, but it's not one he's happy with. He mentions that there are narwhal tusks in the Danish Kunstkammer and suggests that someone compare them to a mammoth tusk. He finishes by saying, "Should any one else hereafter, account better for these Appearances, I shall willingly retract my Opinion."

Strahlenberg did not include an illustration of the mammoth in his book, but it's clear that the whale-like creature he had in mind is nothing like the mammoth we know and nothing like Kagg's drawing, either. In his book, he makes no mention of legs, claws, or fangs. In the nine years between when he left Kagg in Tobolsk and when he published, Strahlenberg had plenty of time to revise his opinions. He mentions having read Müller's account and that of Lorenz Lange, the secretary on Bell's mission to China. However, I think he was not the artist. Kagg's drawing and the description accompanying it do not match. The drawing is of an animal about twice as long as it is tall with claws and fangs and horns fantastically twined around each other. The horns are about half the length of the body. The description says the animal is ten times as long as it is tall. It does not mention claws or fangs, and the horns are slightly less than its height. The entwined nature of the horns isn't mentioned. The drawing bears some resemblance to Müller's description in proportion, and the twisted horns might be an attempt represent the movable horns, which he says can "lay cross-ways over each other."

The written description accompanying the drawing bears some resemblance, in proportion, at least, to that supplied by the "ancient painter," Remessow. Remessow is Semyon Ulianovich Remezov, a life-long resident of Tobolsk, engineer, architect, historian, cartographer, the largest source of information for Strahlenberg's map, and an icon painter. Strahlenberg is being an ungrateful jerk in dismissing him as a mere painter especially since, in the only context that he gives, he is to choosing Remezov's description as the most dependable and using it to base his own conclusions. It's a shame that they were completely wrong. Remezov was never in the place where Strahlenberg says he saw the whale-like skeleton. Although Remezov produced the first atlas of Siberia, he did it by collating observations made by others. The only parts he surveyed himself were along the Ural Mountains. If Remezov was the source of the story, he would have been repeating something told him by one of his own informants, possibly one of his sons who traveled deeper into Siberia than he ever did.

Despite being wrong, Strahlenberg's approach to the mammoth represents positive movement in a new direction. Before his writings, all accounts of the mammoth had been descriptive—observations of the ivory and repetition of the opinions of Russians and Siberian natives—and not analytical. Most writers tried to fit the mammoth into a biblical framework. Strahlenberg was not free from a biblical worldview. In fact, he was a very religious man. His accomplishment was in removing the biblical explanation from its privileged position, rejecting it, and searching for a naturalistic explanation based on several lines of evidence. His discussion of the name of the mammoth revolves around the biblical Behemoth, but he doesn't consider for moment that the mammoth is a Behemoth. His discussion is a matter of historical linguistics only. He considers the Deluge as possible explanation for the presence of the bones and rejects it, though with firm assurances that he is not questioning the historical reality of the Deluge. If he had published sooner, his conclusions might have been of some influence during the decade. But that honor would go to another.

There is a chance that Strahlenberg met Vassili Tatishchev while still in Russia. He arrived in Tobolsk at about the same time that Messerschmidt and Strahlenberg left on their expedition eastward. Even if they did not meet in person at that time, they soon heard of each other and became correspondents. Tatishchev was the ideal civil servant for Peter the Great. He was intelligent, conservative, well versed in multiple fields, and eager to take on new challenges. Peter entrusted him with a number of special projects. In 1720, his was made director of mines and sent to the Urals to locate and develop to new sources of ore. One Swedish prisoner we know for a fact that he met in Siberia was Schönström. The two shared a deep interest in history. Tatishchev helped Schönström locate books and other materials. The friendship might not have been entirely innocent. Schönström was the cousin of Emanuel Swedenborg, the assessor of the Swedish Board of Mines and later a famous mystic. After the war, Tatishchev used Schönström to gain a proper introduction to Swedenborg and to the Swedish Scientific Society. A few months before his death, Tsar Peter gave Tatishchev permission to travel to Sweden and study their mining technology and other industries and engage in a little political spying. Tatishchev stayed for a year and a half.

Tatishchev used his time in Sweden to pursue far more than his official goals. He visited Strahlenberg several times and helped him edit his short history of Russia. In the spring of 1725, he met with Benzelius. At some point, the conversation touched on the mammoth. Tatishchev revealed that during his own travels in Siberia, he had done some research into the subject. At the end of 1721, in his capacity as director of mines, he had sent a detailed questionnaire to officials in various districts. In it, he not only asked about minerals, he also questioned them about other aspects of natural history, antiquities, and subterranean petrifactions. The questionnaire included instructions for specific objects to be sent to him. One of the listed objects was mammoth bones. He later explained that this wasn't just for his own curiosity. The tsar himself had ordered Tatishchev to hunt for a complete mammoth skeleton. Later, Peter authorized him to offer a reward. It would not be claimed in Peter's lifetime, or even in that century. The information provided by Tatishchev thrilled Benzelius. He asked him to write an article for the society. Tatishchev did so in the form of a letter to Benzelius dated May 12, 1725.

The letter is well organized, detailed, and more scientific that anything written before. Tatishchev starts out by describing mammoth ivory and how it's used in trade. He adds a new piece of information here and that is an estimate of the size of the annual trade, fifty to one hundred puds (one to two metric tons). He then describes the beliefs of uneducated Siberians, repeating the Müller/Novitsky account of movable horns, adding that some believe that fresh air that kills mammoths when they tunnel out of riverbanks. This is followed by the "opinions of learned men, which however do very widely differ from each other." Some believe that these are elephants from the East Indies carried north by the Deluge and buried. By now they recognize that the curve of mammoth tusks are different than elephant tusks, but ascribe the difference to their being warped from being buried so long. He mentions the theory of Burnett and others that the the North might have been warm once was thrown into the the cold by a sudden "mutation of the axis of the earth." He mentions the joke of nature idea and finally the possibility that mammoth remains are Behemoths.

His finishes with an account of his own researches and conclusions. Through Tatishchev's official position, he had acquired three relatively complete tusks. Although Peter the Great had died by then, he had not done so before beginning his cabinet of curiosities (the Kunstkammera) and founding a university. He sent to two best tusks to those institutions and kept the third for his own experiments, which, among other things, involved carving "various pieces of work." Much as he tried he was not able to find an intact skull, but he was able to find a rotten and crumbling one to give him a general idea of its size and shape which he prononced elephantine. Tatishchev had more authority than most to make that judgment. He had actually seen an elephant. The shah of Persia sent one to Peter the Great that lived for several years in St. Petersburg. After its death, it's bones were given to the Kunstkammera.

Tatishchev did not dismiss all native beliefs out of hand as simple superstition, he tested one of them. One of the pieces of evidence that Siberian natives offered for their belief that mammoths tunneled through the earth was the rapid appearance of humps and pits in the ground. Such humps and pits, both small and large, are common and mysterious features of permafrost soils. It would be another two hundred years before that was understood. Tatishchev found a newly formed pit and had himself lowered into it. At the bottom, he found moving water, which convinced him that pits were caused by underground rivers. He took water samples and examined them. The water was filled with lime, but the tusks he examined were not. Of the opinions of the learned, he rejects the idea that they are elephants saying all the explanations, like the Deluge or Hannibal, for how elephant bones could end up in Siberia are flawed. To him, all the available evidence revealed was that the mammoth was an unknown animal of unknown nature.

Tatishchev finished the letter humbly: "I, therefore, conclude, that so long as anyone cannot aver, that he has seen this animal, many doubts must still necessarily remain, which must be left to time and farther observations to clear up." Though the letter offered no solution to the mammoth question, the society seemed happy with it. Within months they published it in the form of a small pamphlet. The letter was also published in an English language newspaper, adding to its circulation. Four years later they published it again in a slightly revised form in their journal. After that, the mammoth dropped off the agenda of the society. They had questioned the returning prisoners, examined all the bones and teeth they could, and they had sought out expert testimony from someone well placed gather evidence. There was little else they could do at that point.

But the society didn't have the last word out of Sweden on the topic. As the mammoth question was winding down, Olaus Rudbeck took in a brilliant student as a boarder. Karl Linné (Carolus Linnaeus) would soon become famous for his scheme for cataloging and naming everything in the natural world. In the first edition of his *Systema Naturae* (1835), the mammoth, in the form "Mammatowacost," appears in small Gothic script as the last entry in the mineral kingdom. Linné did not mean to imply that mammoths were a mineral product; his decision to place it there was based on the same logic that caused Ludolf to call it a mineral, the old definition of fossils as things found in the earth. Linné had a whole subcategory in the mineral kingdom for "Petrified Quadrupeds." It was here that he placed mammoths. He stubbornly kept mammoths and other fossils in the mineral kingdom for thirty years before finally dropping the entire mineral kingdom from his classification scheme. Later, when asked about the mammoth, he imperiously pronounced that it was a giant walrus and that was that.

Despite the intellectual exchanges that took place during this sorry chapter in Sweden's history, the fate of the prisoners after Poltava was a tragedy, pure and simple. Thousands did not survive their captivity. Those who did survive were separated from their families and communities for more than ten years. They came home as strangers. But it is undeniable that the officers' time in captivity produced a treasure trove of scientific progress. Not only did they advance European knowledge of the archaeology, anthropology, geology, geography, and natural history of Russia and northern Asia, they trained a generation of Russian scientists and provided them with intellectual contacts in the West. From this, can we come to the Panglossian conclusion that everything works out for the best? Of course not. The world would not have been any worse off if it had taken us twenty years longer to learn about Siberia and more prisoners had survived.

CHAPTER 5

ACADEMIES AND JOURNALS

Between the villages of Burgtonna and Gräfentonna, near Erfurt in Thuringia, is a hill, and beneath that hill is a layer of clean white sand that has proven useful in several handicrafts, including filling hourglasses. Because of the value of the fine sand, workers in the quarry are careful and methodical in harvesting the sand so as not to contaminate it. In December 1695, they uncovered "some awful big bones" in the sand and, in a scenario now familiar, sent word to the castle to find out what to do with them. Luckily for us, the lord of the land, Duke Fredrick II of Saxe-Gotha-Altenburg, was an enlightened despot who was both a patron of the arts and sciences and an avid collector. More than simply ordering the workmen to save the bones for his collections, he had them leave them in place and slowly uncover them. There is no record of who directed the excavations, but several educated people kept tabs on the work, including the duke's librarian and Witsen's loyal correspondent Wilhelm Ernst

Tentzel. The bones that the workmen had uncovered were revealed to be those of a leg. The foot had five digits and a short ankle. Many observers thought they looked close enough to human to believe the workmen had uncovered a giant. After the lower portion of the limb was uncovered, some nasty weather postponed the work for a few weeks. When the workers returned in January, they proceeded up the body, uncovering the upper leg bones, a pelvis, vertebrae with ribs in place, shoulder blades, forelegs, and then a "hideous head" unlike anything they had ever seen. Here, they were instructed to excavate a large space around the skull so that it could be viewed from all directions. Lying near the top of the skull were two eight-foot-long ivory tusks. With the entire skeleton nicely uncovered, the duke made a special trip from Gotha on January 23 to view it, bringing along a large entourage that included a number of doctors from the university and Tentzel.

Tentzel and the doctors, led by Johann Christoph Schnetter, all had a good laugh over the silly peasants who thought the bones were those of a giant. This much had changed in the eighty years since the Theutobochus controversy. By the last decade of the century, the numbers of those who still believed that there had ever been giants, other than the few individuals named in the Bible, were rapidly shrinking. There would be holdouts well into the next century, but giants were rapidly becoming irrelevant to scientific debates. While the doctors and Tentzel agreed on what the bones were not, they passionately disagreed about what they were. Schnetter and the doctors believed they were mineral formations, *unicornu fossili*, while Tentzel believed that they were the remains of a real elephant. Duke Fredrick chose not to take sides. He ordered the doctors and Tentzel to each submit a brief summarizing their arguments.

The doctors organized their arguments, and Schnetter wrote them up and had them published as a pamphlet and distributed around Europe by St. Valentine's day. The entire pamphlet is seven pages long, and a sizable chunk of it is dedicated to describing the discovery. The title contains their conclusions: *Kurtze doch ausführliche Beschreibung Des Unicornu Fossilis, oder gegrabenen Einhorns, Welches in der Herschafft Tonna gefunden worden* (A short but detailed Description of the Unicornu Fossilis, or excavated Unicorn, which has been found in the Lordship of Tonna). Schnetter spends very little space laying out the argument itself. He assumes that most of the audience is already familiar with the idea of *unicornu fossili*. The largest part of the pamphlet is dedicated to citing contemporary writers who describe discoveries of *unicornu fossili* in neighboring parts of Germany. One piece found near Tonna a few years earlier resembled a stag's horn, but was true *unicornu fossili*. When administered to patients, it cured them of "the falling sickness." This presentation is another significant change from the arguments of Habicot and Riolan. Schnetter makes a few references to writers of antiquity, but most of his citations are to his contemporaries and discoveries made in recent years. Between these two parts, he makes a preemptive strike against Tentzel by explaining why the supposed bones could not be an elephant. One point he makes is that while the bones are not scattered, they are somewhat disarticulated. Each bone is separated from the next at least by the thickness of a hand. Another point he makes that the tusks appear to be hollow, not solid ivory. Schnetter could not have known that elephant tusks, as opposed to walrus tusks or narwhal teeth, are hollow for almost half their length. This fact had not been mentioned in any of the accounts available to him. While he admits that the skull bears some resemblance to an elephant's, he writes that it cannot be one because the tusks in question are up by the eyes, and not by the mouth, where everyone knows an elephant's should be. The doctors may have had a vested interest in this solution. Though prices had dropped steadily all through the century, that much unicorn horn would have brought a pretty penny in the medicinal market.

Tentzel wrote a short response, which he submitted directly to the duke. He took more time writing a detailed statement of his case and, by taking more time, was able to prepare a full rebuttal to the doctors' arguments. He had a special advantage in preparing his case. As curator of the duke's collections, he had access to fossils and other curiosities that he could compare with the bones. He had the bones themselves; the duke had had him collect as many of the remains as he could. By taking more time, he was able to interview the diggers and other witnesses to excavation. And he had Mullen's pamphlet, with its detailed drawings of the skeleton and skull of the Dublin elephant. Tentzel's public presentation appeared in the April issue of his journal *Monatliche Unterredungen einiger guten Freunde von Allerhand Büchern* (Monthly Conversations between Good Friends about All Kinds of Books). It runs 108 pages with an illustration of the skull. The skull almost exactly matches the one in Mullen's pamphlet. To emphasize the similarities, Tentzel shows his skull from exactly the same angle as Mullen showed his. After a detailed description of the discovery, the fictional friends of the title take sides. Caecilius and Passagirer take Tentzel's position, and Aurelius and Didius defend the doctors'. Naturally, most of the space is given to the former.

This was not the first time Tentzel had written about elephants or buried ivory in *Monatliche*. He had already given thought to the issue and developed opinions on the subject. Like Habicot eighty years before him, Tentzel examines not just the form of the bones, but also their texture, declaring that they are too complex to be "sports of nature," Stones which only coincidentally resemble bones. A large part of his treatise on the elephant is dedicated to making this argument. Next, armed with Mullen's drawings he examines the individual bones finding that, although they are much larger, they almost exactly match the Dublin elephant in shape and proportion. Among other observers at the site was a Dutch sailor who had spent many years in India. When Tentzel interviewed him, the sailor informed him that elephants keep growing for their entire lives. By the size of the tusks, he estimated that the Tonna elephant must have been at least two hundred years old. (Overestimations of the life expectancy of elephants were typical in the literature of the times. Accurate knowledge of their life expectancy—fifty to seventy years—wouldn't be available in Europe until the beginning of the nineteenth century.) Finally, Tentzel consulted the work of the pioneer microscopist Anton van Leeuwenhoek, who had written about the internal structure of ivory, to determine that the ivory was real elephant's ivory. He even traveled to some neighboring principalities to examine ivory and bones in other princely collections.

Once he had determined that the bones and ivory were real organic remains and that they belonged to an elephant, Tentzel attacked the problem of how the skeleton of an elephant came to be buried in central Germany. He considered only two possibilities: that someone had brought the elephant there when it was alive, or that it had floated there during the Deluge after it had drowned. He did not consider the possibility that elephants had once been native to Germany. Tentzel was aware of other bones and teeth found around Europe. He mentions specifically Boccaccio's Sicilian giant, the Krems teeth, and the Theutobochus bones. In his opinion, they were all almost certainly from elephants. He understands that the problem isn't just this one elephant; it is that there are the remains of many elephants spread across Germany and other parts of Europe. To those who thought that the elephants had been brought north by the Romans or the Huns or had been the famous elephants owned Charlemagne, Henry II, or others, he had one question: Why did they leave the ivory behind? "The use of ivory," he wrote, "which is very ancient, will not allow us to believe that upon burying the elephant there they should neglect to take away the teeth, which are very large and fine." He followed this by saying he found it hard to believe that anyone would dig a pit twenty-four feet deep to bury their dead elephant. This observation introduced his treatment of the Deluge. Tentzel carefully examined the strata of soil around the skeleton and determined that the layers above had never been disturbed. Even if the owner of the elephant had dug a twenty-four-foot pit to bury it, leaving behind the ivory, the soil above the elephant would have been mixed together. What he observed were seven undisturbed strata above the white sand where it lay. The primary mechanism for laying down strata that Tentzel was aware of was sedimentation beneath bodies of water. He was familiar with Steno's work, but rejected the concept of deep time. He preferred to believe that the strata had been all laid down at once. When Aurelius and Didius get their turn to speak, to Tentzel's credit, they give an accurate summary of the doctors' position rather than a rude parody of it. Nevertheless, they still lose the debate.

One of the readers of the *Monatliche* article (and possibly a subscriber) was Leibniz. Soon after the issue came out he wrote to Tentzel to express his qualified support. He begins by saying he completely agrees that the doctors are wrong. In writing his *Protogaea*, Leibniz had studied fossil shells and bones and come to the conclusion that the great majority of them are the real remains of organic creatures. He categorically denies that the Tonna bones could be jokes of nature. His reservations are with Tentzel's conclusion that his bones come from an elephant. He agrees that the climate in Germany had never been suitable for elephants. His preferred explanation is that the bones had once belonged to some kind of unknown marine animal "similar to the elephant." The same illustration he used to show the Quedlinburg unicorn included a mammoth's tooth labeled "Tooth of a marine animal." Most of the fossils that Leibniz features in the *Protogaea* are seashells. They are found in all of the parts of northern Germany and are especially common in the Netherlands. From this he concludes that that Thuringia had once been the northern coast of Germany and home to walruses. In the many centuries since the Deluge, northern Germany had been slowly covered by soil brought down by rivers from the Alps and other mountains of southern Germany. Three weeks later, Tentzel wrote back to say that he is still confident that the bones are those of an elephant. To support this, he points out that elephants and walruses have very different bodies and that the Tonna bones were of a creature many times larger than a walrus.

Though Tentzel was confident in his conclusion that the bones belonged to an elephant, it didn't stop him from hunting for additional confirmation. He knew that there was another complete elephant's skeleton he could compare with the Tonna bones. That elephant, named Hansken, was quite famous. After the Dutch East India Company beat the Portuguese out of the India trade, one of their agents acquired a young female elephant in Ceylon in 1637. Once in Europe, her owners taught her to wave a flag, beat a drum, manipulate small objects, and other tricks, and sent her on a tour of the continent, where she performed before paying audiences. After eighteen years on the road, she injured her foot in Italy, developed an infection, and died in Florence on November 9, 1655. Grand Duke Ferdinando II had a special mass written for her. Ferdinando II was tutored as a child by Galileo and grew to be a patron of the sciences. He had most of the good parts of Hansken removed before burying her. He had the skeleton mounted as accurately as possible and had her skin stuffed with straw for display in his collection. In July, Tentzel wrote a letter to Antonio Magliabechi, the personal librarian to Cardinal Leopoldo de Medici, the brother of the grand duke. Magliabechi was one of the major figures of the Republic of Letters during that generation and was widely renowned for his erudition, encyclopedic memory, and disgusting personal hygiene. He had been a friend of Steno's and was familiar with his theories about the organic origin fossils. In his letter, Tentzel repeated most what he had written in *Monatliche Unterredungen*, leaving out the literary flourishes and defense of the doctors' position. Tentzel wanted Magliabechi to approve his interpretation of the bones and hoped he would send a detailed description of Hansken's bones and loan him one or two to compare with those in his possession. Magliabechi enthusiastically endorsed Tentzel's conclusions and sent the detailed description he requested, but no bones. Italian scholars, as opposed to those north of the Alps, had no trouble accepting the presence of elephants on their lands. Historical records told of many elephants that had been in Italy during Roman times. First, there were the war elephants of Pyrrus and Hannibal. Later, there were the large numbers of elephants brought by the Romans to be slaughtered in the circuses for entertainment. Fossil elephants regularly turned up and, thanks to Hansken, were easily recognized as such, though always credited to historical times. Magliabechi forwarded Tentzel's letter to several other Italian thinkers, who also agreed with his conclusions. They wrote their own pamphlets and letters to journals in support. One, Paolo Boccone, who was familiar with Hansken's skeleton, made a trip to Gotha to examine the Tonna bones.

After writing to Magliabechi, Tentzel and Leibniz engaged in a new round of correspondence on the subject that lasted for the rest of the summer. In his first letters, Leibniz concedes that an elephant is a more likely source for the bones than a sea creature, but says that a sea creature is still his preferred solution. The sea, he writes, produces animals, such as whales, that are much larger than land animals. He then goes on to outline his theories of the ancient earth. Referring to his belief that northern Germany was underwater in the post-Deluge era, he says that "the Earth has undoubtedly undergone greater changes than the common man thinks." The Deluge is not necessarily the only possible explanation for the presence of elephant bones at Tonna. He offers two additional explanations: Germany might once have been warmer than is is today, or there might once have been elephants that were adapted to a colder climate. He's vague about when this time might have been. For the former, he refers to Thomas Burnett's theory that the earth's axis might once have been perpendicular to the ecliptic rather than tilted as it is today. This would have made all the earth comfortably warm, rather than hot at the equator and frozen at the poles. In this climate, normal elephants could have flourished in Germany. He leaves out the inconvenient detail that Burnett thought that the catastrophe responsible for the shift in the earth's axis also caused the Deluge. Of the two theories, Leibniz thinks the second is more likely; there might once have been elephants that were adapted to Germany's climate, an idea he had rejected a few months earlier. He points out that there are many species in the new world that are unknown in the old. By his own theory that there have been many smaller, regional catastrophes since the Deluge, it makes sense that there have been many regional extinctions. How many species must have been lost when Atlantis sank, he asks?

In a letter written in August, Leibniz is careful to assure Tentzel that, when he says species were lost, he is not claiming that the species are totally extinct: "I think we must distinguish between extinct species and those that have greatly changed." At the time that they corresponded, total extinction was an unthinkable, even heretical, idea. Extinction implied a flaw in God's creation. Since Aristotle, the dominant philosophies in Europe had treated all that existed as a ranked hierarchy known as the *scala naturæ* (ladder of nature) or great chain of being. At the top sat God. At the bottom was common dirt. In between were angels and demons, celestial bodies, humans (ranked according to class), land animals, fish, plants, and minerals. There was a place for everything and everything was in its place. For the chain to be perfect, every part must have a role to play in maintaining a harmonious whole. Alexander Pope, in his "Essay on Man," described this problem:

> Vast chain of being! which from God began,
> Natures ethereal, human, angel, man,
> Beast, bird, fish, insect, what no eye can see,
> No glass can reach; from Infinite to thee,
> From thee to nothing. On superior powers
> Were we to press, inferior might on ours:
> Or in the full creation leave a void,
> Where, one step broken, the great scale's destroyed:
> From Nature's chain whatever link you strike,
> Tenth or ten thousandth, breaks the chain alike.

If links could be removed without destroying the whole chain, they must have been superfluous in the first place. It was inconceivable that God would create things with no purpose, and thus the issue with extinction in this worldview.

In the variation of the *scala naturæ* that Leibniz subscribed to, ladders or chains were not the best metaphors. These implied an abrupt jump from one species to the next. His school envisioned each species flowing into the next by way of different breeds and hybrids, a principle he called *lex continui*. Tentzel and Leibniz's correspondent John Ray summarized the idea: "Nature, as the saying goes, makes no jumps and passes from extreme to extreme only through a mean. She always produces species intermediate between higher and lower types, species of doubtful classification linking one type with another and having something common with both." Leibniz had a very broad concept of species. To him, a large group of animals that shared an "inner nature" could all be considered a single species. Following his statement that remains that appear to be extinct were really animals that had greatly changed, he gave examples: "Thus he dog and wolf, cat and tiger can be seen as being of the same species. The same can be said about the amphibious animals or marine oxen [walruses] once analogous to the elephant." Forms could come and go without the species itself going extinct. A cold-adapted elephant had the same relation to a tropical elephant as a shepherd dog to a terrier.

Leibniz's idea that a species could change and adapt required a certain amount of time that was not obvious in the traditional biblical chronology. His belief that the North German Plain had once been a sea that had gradually filled with silt from the eroding Alps also required time. Leibniz and Tenzel were both familiar with an author who had begun to address the issue of time and change. Niels Steensen, better known to the English-speaking world as Steno, has a good claim to be one of the founders of both geology and paleontology. He wouldn't have thought so, if for no other reason than that neither word had been coined yet. Steno was trained as a doctor and had a strong interest in anatomy. His search for an ideal position led him from his home in Denmark to Tuscany, where Grand Duke Ferdinando II, the owner of Hansken's remains, hired him as his personal physician in 1666.

Several months after Steno arrived at the court, fishermen in Livorno landed a huge shark. The grand duke heard about it and ordered its head brought to Florence so Steno could examine it. The head was getting rather ripe when he began his dissection, but this did not prevent a large crowd of academy members and courtiers from gathering to watch. Steno started on the top of the head, examining the skin and brain, which he found surprisingly small, before rolling it over to look at the mouth. In examining them he was struck by the resemblance between the teeth and glossopetrae, or "tongue stones," as they were known. These distinctive triangular stones were found in many parts of Europe, but were especially common in Malta. Various traditions likened them to the tongues of serpents or birds. One tradition said they fell from the moon. Another said they were real tongues of snakes that had been petrified as punishment for annoying a saint. Like unicorn horn, they were thought to be effective in treating a variety of illnesses. Steno was not the first to notice the resemblance. However, he was the first to strongly argue a theory about their origin. Steno categorically rejected the theory that glossopetrae could grow inside solid rock influenced by cosmic or any other forces. He compared them to crystals, which did grow in the earth. Crystal growth could be re-created in the laboratory. Close examination showed that crystals had very simple structures made up of elementary forms repeated. Glossopetrae, however, had complex structures unlike any known type of crystal. His theory reversed the order of events. He argued that the glossopetrae were not objects resembling tongues that somehow formed within rocks, but real shark's teeth, and that the rock in which they were found had formed around the teeth after the sharks died.

Steno's idea of stones forming addressed an element of geological time that was not yet part of the European intellectual tradition in 1670. In a literal reading of Genesis, the earth was only three days older than Adam, making the idea of earth history as different and separate from human history rather pointless. This did not mean that the world was not unchanging. Recorded history produced accounts of some changes. Volcanoes erupted, earthquakes shook the earth, and rivers ate away at their banks and filled harbors with silt. These changes were trivial compared to the two great events that formed the basic shape of the world: Creation and the Deluge.

Many writers before Steno had commented on the stratification of stone and soil on the earth. Understanding them was simple: they were laid down as sediment during the two formative events. Steno had no problem with this idea until, one day in Tuscany, he realized that some strata were tilted at angles that sedimentation could not produce. His solution to this problem was ingenious. He envisioned one set of strata being laid down, followed by a period of subterranean waters washing away the lower layers until the upper layers collapsed, creating the angled strata that puzzled him. In time, new strata were laid over the angled strata in places, eventually undermined, and collapsed, creating the complex structures he saw in Tuscany. It was during periods of new strata being laid down that glossopetrae were deposited in fresh mud before it hardened into rock. Steno did not explicitly state that this progressive creation of the landscape could take much more time than the traditional biblical chronology, but others would not avoid it.

Tentzel went to great lengths to advertise the Tonna discovery and gather support for his position. Soon after writing his letter to Magliabechi, he had it published as a pamphlet in Latin and sent copies to various other influential thinkers and scientific societies. A few months later, he published an edition of the letter in German. At the end of the year, he printed a second edition of the Latin version. These were generally well received. Abridged versions were published in the Italian *Giornale de'Letterati*, the French *Journal des sçavans*, and the German *Acta Eruditorum*. In London, the Royal Society published the entire Latin version of the pamphlet in their *Philosophical Transactions*. Unlike Grand Duke Ferdinando, Duke Frederick allowed his librarian to give away several of the bones themselves to make his case. Soon after their discovery, Frederick allowed the bones to be shown at the spring fair in Leipzig, where Tentzel showed them to several traveling scholars. Frederick may have had his own motives for giving his librarian such free rein. Besides wanting to be known as a patron of the sciences, he had just two months before the discovery been awarded the highly prestigious Order of the Elephant by the king of Denmark. It didn't hurt to remind people of this fact by having his name associated with elephants. Tentzel wrote a follow-up piece in the January 1697 issue of his *Monatliche*, telling his subscribers about the support he had received and other accounts of buried elephant bones that had been brought to his attention. Here he mentions the account in Witsen's shipbuilding book account of elephant ivory discovered in Mexico and Siberia, using Witsen's name for the latter: *Mammotekoos.* He follows this with a direct quote from Ludolph's grammar, published the previous year, calling the ivory *Mammotovoi kost.* The support that Tentzel received from other scholars did not cause Schnetter and the doctors to gave up on the debate. The medical college published at least four more pamphlets defending their position over the next year. Presumably Schnetter was the anonymous author of all four. Tentzel answered each in due course, but neither side said anything new in these pamphlets.

After Magliabechi, Tentzel was most eager to get the approval of the Royal Society. The Royal Society (officially the Royal Society of London for Improving Natural Knowledge) was the first of the scientific societies and academies that began to appear in the middle of the seventeenth century. These societies began as informal salons of curious men near centers of learning and rapidly began to gain official, state-sanctioned status beginning in the 1660s. The Royal Society began at Gresham College in London in 1662, the Académie des sciences in Paris in 1666, and the Swedish Royal Society of Sciences at Uppsala University outside Stockholm in 1710. As Russia had very few native scientists, Peter the Great recruited some from Western Europe to move to St. Petersburg and create the Russian Imperial Academy in 1724. Other groups appeared in the Italian and German states. Parallel to the rise of the societies was the rise of academic journals. The *Giornale de'Letterati,* founded in Rome in 1668, included studies of antiquities and church history along with reports on scientific advances. The French *Journal des sçavans,* whose first issue came out in January 1665, specialized in book reviews and obituaries. Prior to the 1660s, the only forms of scientific writing had been letters, books, and casual notices in popular gazettes. The journals making their debut by the middle of the century were different in that they drew more directly from the researchers themselves. The topics that each journal covered were defined by the interests of the members of the group and limited by the local church and royal censors. Some journals contained articles about history, philosophy, and (rarely) theology. Others leaned almost exclusively toward the sciences. As this was the pioneering era of the academic journal, the styles and formats varied wildly and only remotely resembled their modern descendants. The most common contents were the minutes of presentations at the meetings of their parent organizations, letters from corresponding members, and reviews of books. At first, the reviews were the most important. Some reviews were simple summaries of works. Others were long enough to qualify as plagiarism under later laws. The most valuable were critical essays akin to the best modern reviews. These early journals expanded on the Republic of Letters, but did not replace it. Letters remained important because they were less subject to censorship than published documents. As far as reviews were concerned, journals allowed a much bigger audience to become aware of interesting works and seek them out. It was the pursuit of a specific audience that led Tentzel to seek out the Royal Society.

The Royal Society did not have a journal when it was chartered. The decision to publish one came three years later when the fellows asked the society's secretary, Henry Oldenburg, to edit and print a monthly account of the society's communications. Oldenburg was allowed to keep any profits from the venture as a form of payment for his troubles. *The Philosophical Transactions of the Royal Society* began as sixteen pages of short notices. The first issue has a one-paragraph description of a spot in Jupiter's atmosphere reported by Giovanni Cassini and a number of notices from Robert Boyle concerning topics as diverse as lead ore in Germany, whaling off Bermuda, and "An Account of a Very Odd Monstrous Calf." By the seventh issue, there was the first account of fossils, some petrified wood, and perhaps a bone. For the first year, the *Philosophical Transactions* consisted mostly of short excerpts from letters to Oldenburg. This began to change during the second year as he began to include longer pieces describing experiments and observations of nature done by society members and their correspondents. The emphasis on experiments and observations soon gained the Transactions a reputation for rigor and a place of prestige among journals. Despite the fact that most pieces in the journal were in English—a language few continentials spoke—publication in the *Transactions* was considered a very desirable sign of approval. Each issue of *Journal des sçavans* usually included a summary of one or more articles from the *Transactions.*

Although the *Transactions* published his pamphlet in full, Tentzel was disappointed with the society's response. He had hoped for some discussion or commentary among the members or private discussions through letters. Instead, it was published with a short note saying the bones and ivory he sent agreed with his description and that they had been placed in the society's repository. He wrote to John Ray, a fellow of the society, several times and even sent him some bones, but never received the kind of engagement he hoped for. Despite their silence, the members of the society remembered his pamphlet and bones and would refer to them whenever they discussed elephantine bones for the next several decades.

As these societies and journals grew in prominence and influence into the eighteenth century, another German discovery, less than five years after the Tonna bones, garnered attention due to the sheer magnitude of the find. In late April 1700, a soldier doing guard duty outside Cannstatt in Württemberg noticed some ivory protruding from the ground near an ancient wall. There are no descriptions of the discovery that match the detail of Tentzel's, but the duke's personal physician, Solomon Reisel, did write a short letter covering most of the pertinent details. Duke Eberhard Ludwig was not particularly interested in affairs of state, preferring affairs with his mistress. Luckily for us, he was another enthusiastic collector. The excavations he ordered lasted a full six months. During that time, an enormous amount of bone and ivory was uncovered. In some places, they were packed so tightly that the workers had to blast them loose. By the time the digging stopped, they had recovered sixty tusks, mostly in pieces, and the bones of bears, wolves, cattle, various small animals, and many mice. The inclusion of the last suggests that while the excavation was violent at times, it was very thorough. The duke picked out several nice bones, teeth, and pieces of ivory for his collection. The rest he gave to the apothecaries in Stuttgart to grind up and sell as unicorn horn.

Reisel and David Spleiss, a doctor from Schaffhousen, Switzerland, who also wrote about the bones, were of the opinion that the ruined walls were the last vestiges of an old pagan temple. The bones, they believed, were the remains of sacrifices conducted there. Roman artifacts had been found in higher strata nearby. Reisel was unsure whether the largest—mammoth—bones were real and not sports of nature. Spleiss had no doubt that they

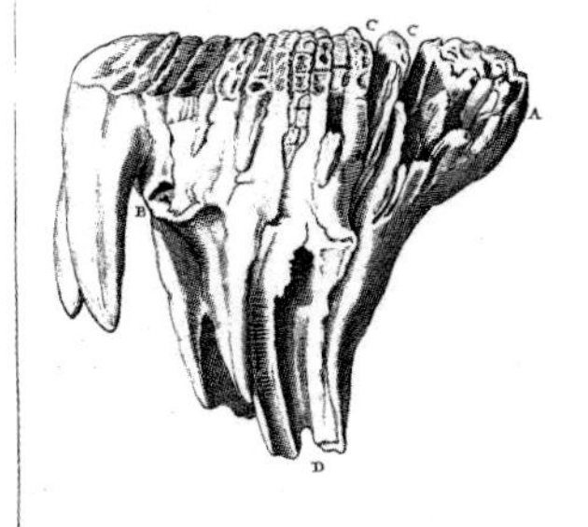

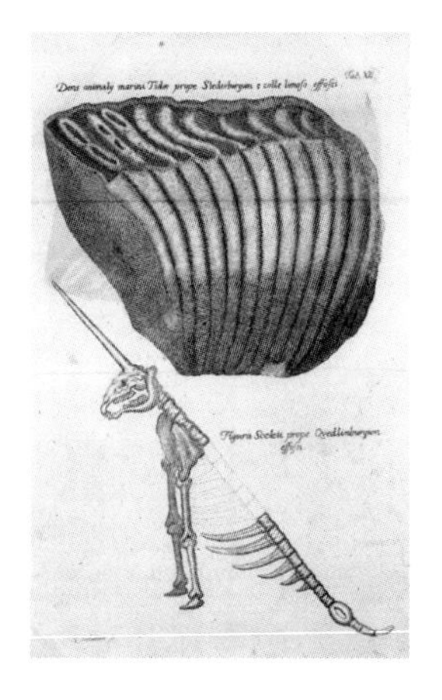

FIGURE 1 (TOP): *The giant's tooth found near Krems, drawn from life by Matthew Merian (1651).*

FIGURE 2 (BOTTOM LEFT): *The relative sizes of famous giants according to Athanasius Kircher (1678). From left to right: The giant mentioned by Boccaccio, Goliath, the Leyden giant, and four classical giants described by Pliny.*

FIGURE 3 (BOTTOM RIGHT): *The Quedlinburg unicorn. The earliest account is a secondhand retelling by Otto von Guericke in 1672. The earliest surviving illustration didn't appear until 1704.*

FIGURE 4 (TOP): *The white elephant (*Ylefanz blanc*) of Rucheni, located in northwestern Russia on a 1550 map by Pierre Descleliers.*

FIGURE 5 (BOTTOM): *Walrus or mammoth? The* morsus *located near Norway on Martin Waldseemüller's 1516 Carte Marina.*

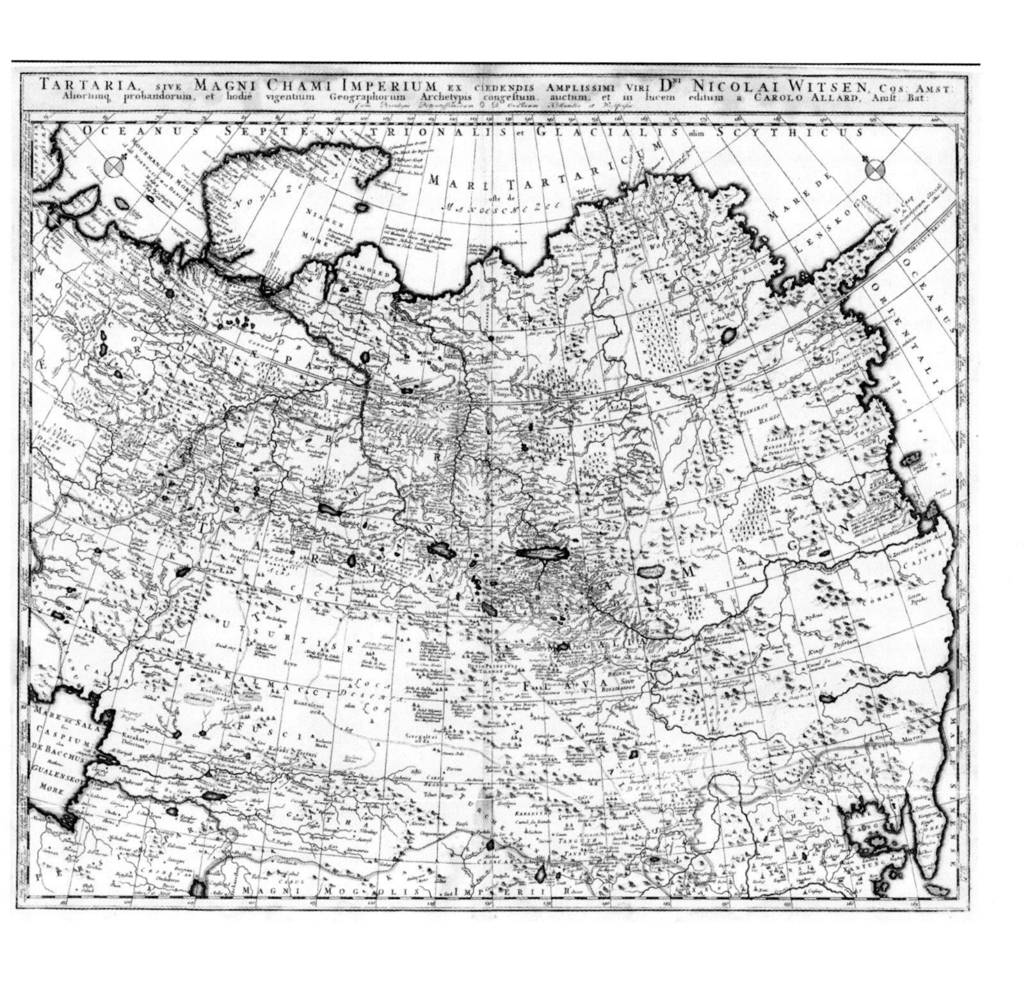

FIGURE 6: *Nicolaas Witsen's map of northeast Asia represented the most up-to-date knowledge of Siberia in 1690.*

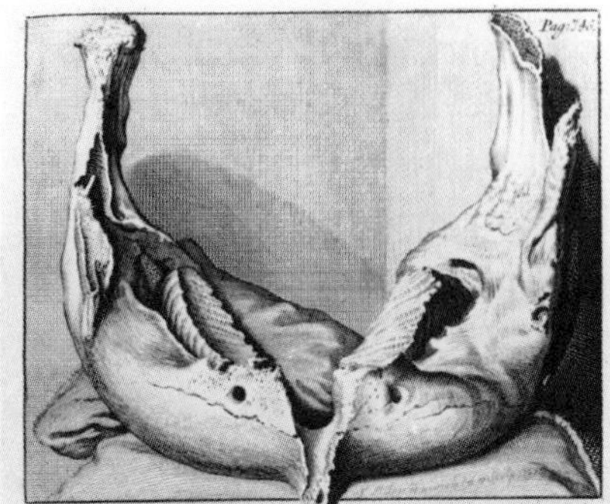

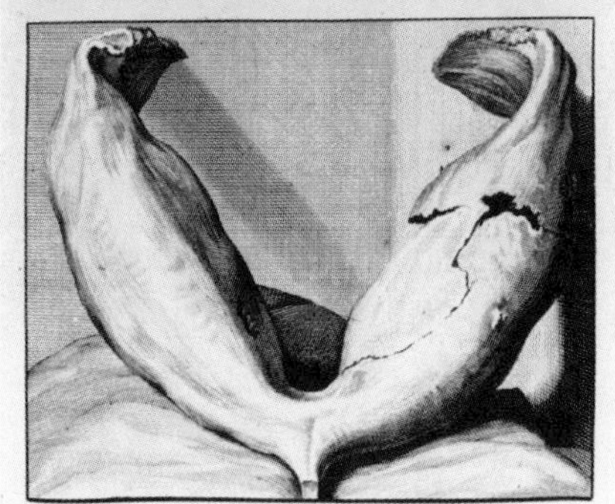

FIGURE 7 (TOP): *A complete mammoth's jaw sent to Witsen in 1707 by his Russian friends.*

FIGURE 8 (BOTTOM): *Baron Kagg's Behemoth (1723). One Swedish POW's idea of the mammoth based on native tales, some bones, and lots of imagination.*

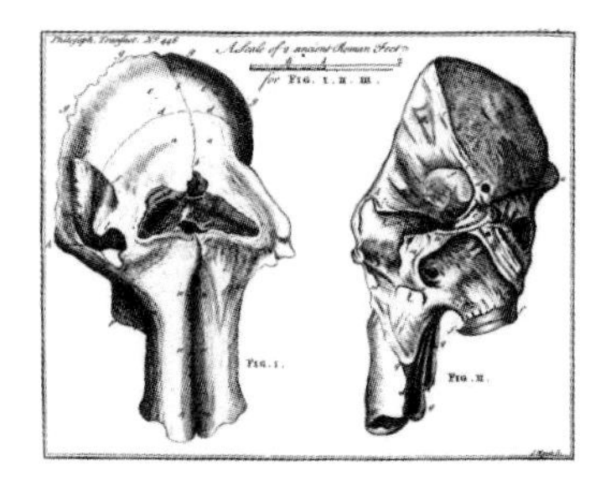

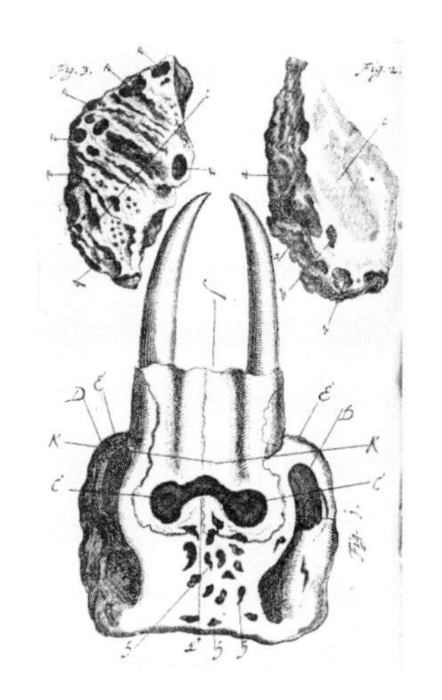

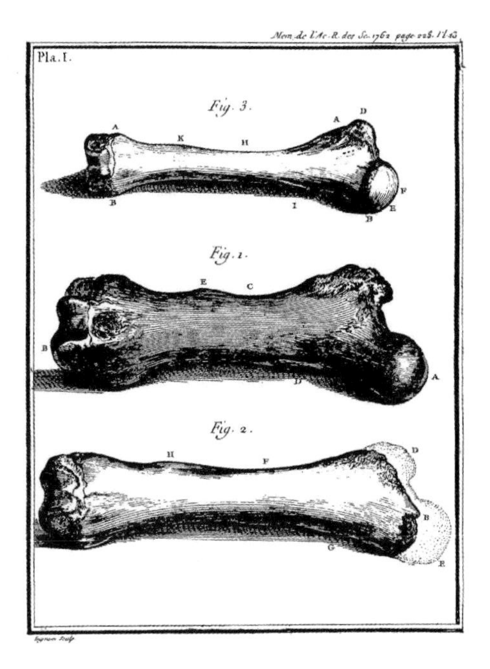

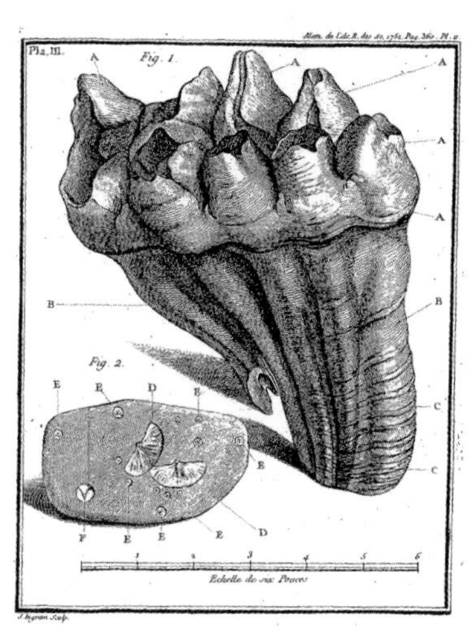

FIGURE 9 (TOP LEFT): *A mammoth skull based on sketches by Messerschmidt. It is the first truly accurate drawing of a mammoth skull to be printed (1738).*

FIGURE 10 (TOP RIGHT): *The skull of an extinct elephant found near Tonna in 1695. The tusks look short because the artist is looking directly at the elephant's face. The written description describes the tusks as almost twice as long as the skull.*

FIGURE 11 (BOTTOM LEFT): *From top to bottom: the femurs of an Asian elephant, a mastodon, and a mammoth. Louis-Jean-Marie Daubenton decided that they all came from the same species, calling the differences in girth nothing more than differences in age and sex (1762).*

FIGURE 12 (BOTTOM RIGHT): *A mastodon's tooth gifted to Jean Etienne Guettard (1752). For a decade, this would be the best image that European savants had to work with.*

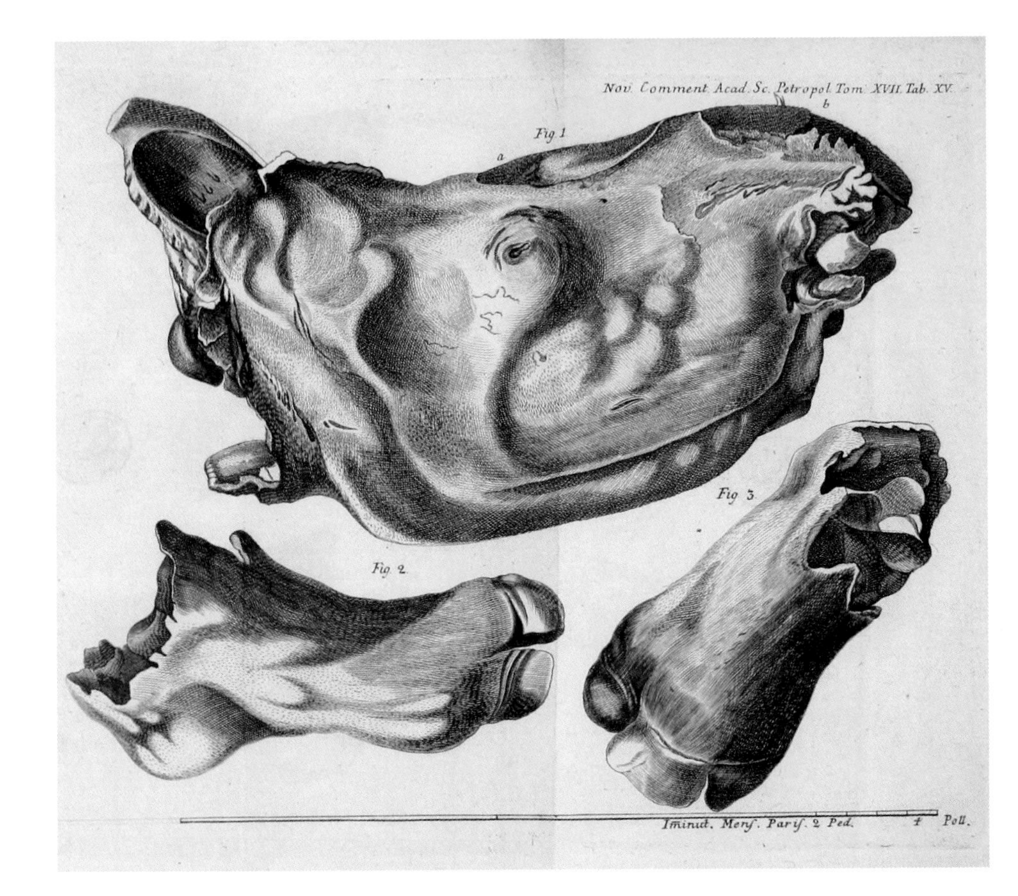
Nov. Comment. Acad. Sc. Petropol. Tom. XVII. Tab. XV.
Fig. 1
a
b
Fig. 2
Fig. 3
Iminut. Mens. Paris. 2 Ped.
4
Poll.

Mammouth.
a.
d
b
e
Elephas primigenius

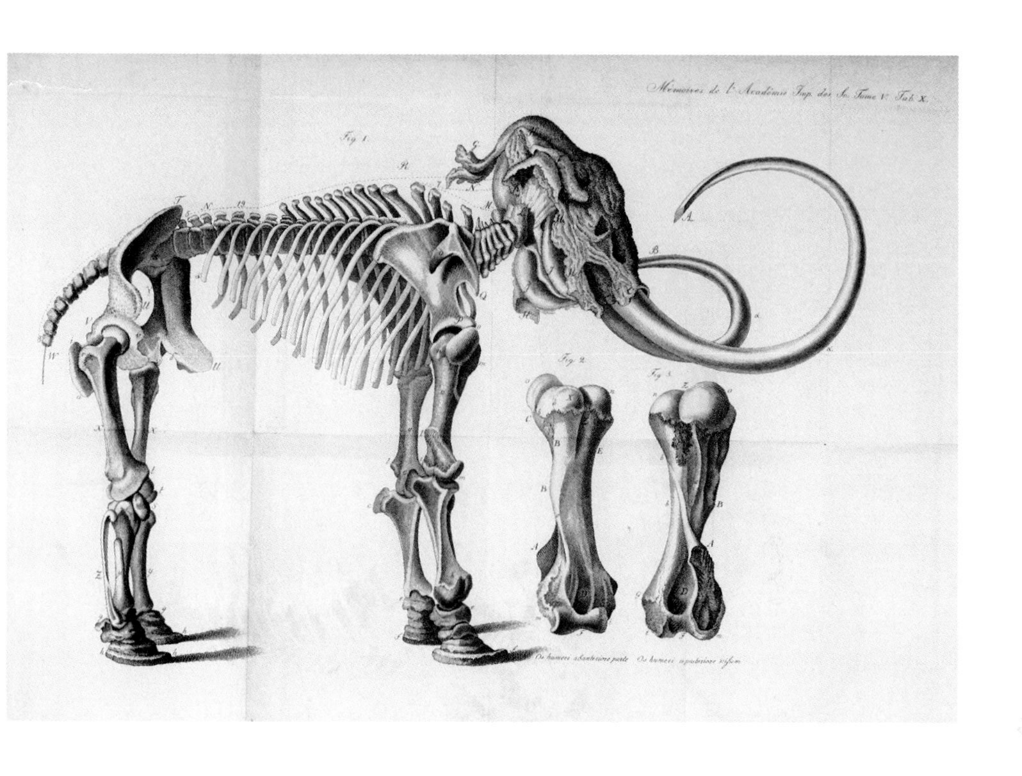

FIGURE 13 (OPPOSITE TOP): *Despite many reports of bloody mammoths eroding out of riverbanks in Siberia, the first frozen giant to be recovered with soft parts such as skin was a woolly rhinoceros found on the Vilui River near the Arctic circle (1772).*

FIGURE 14 (OPPOSITE BOTTOM): *Roman Boltunov's attempt to reconstruct a well scavenged mammoth that he saw in a snowstorm (1805). This copy and a piece of the mammoth's skin were sent to Johann Blumenbach.*

FIGURE 15 (ABOVE): *Wilhelm Gottlieb Tilesius' masterful reconstruction of the first complete mammoth to be recovered (1815).*

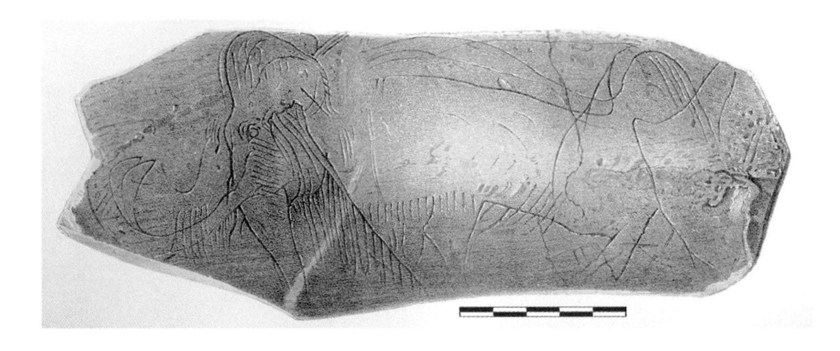

FIGURE 16 (TOP): *The Adams mammoth on display in the Kunstkammera. Tilesius didn't get everything right.*

FIGURE 17 (BOTTOM): *The 1864 discovery at la Madeleine rock shelter of a piece of tusk engraved with the image of a mammoth proved humans and mammoths had lived together.*

were the bones and tusks of elephants. Like Tentzel, who he cites, he closely examined the bones and was certain that no mineral process could mimic the internal structure of real bones. His theory was that the Romans had made an unrecorded military expedition into that part of Germany using war elephants and had been defeated. The victorious pagan Germans had then sacrificed the captured elephants to their gods. Reisel was a correspondent of the Royal Society. No notice of the discovery appeared in the *Transactions* at the time, but the members knew of it and mentioned it along with Tenzel's elephant it in later years.

The first papers published in the *Transactions* to deal with the problem of these massive bones were by Thomas Molyneux, one of a pair of scientifically minded brothers. The older of the two, William, was one of the founders of the Dublin Philosophical Society and both were made fellows of the Royal Society in the 1680s. His first comments were two letters written over the winter of 1684–1685. These "Concern[ed] a Prodigious Os Frontis in the Medicine School at Leyden." He writes that this forehead bone is twice the size of a normal one and must have come from a man eleven or twelve feet high. No one at the medical school knew where it came from. In a much longer article written in 1700, he makes clear that he believes this is merely the bone of an unusually tall man, nearing the maximum tallest possible for a human, and not a giant in the mythical sense. Molyneux's interest in large bones was not limited to human-looking bones. In 1697, he wrote about giant antlers often found in Ireland that resembled those of the American moose. From the illustration, we can tell he was looking at the skull and antlers of the extinct Irish elk. He uses the word "extinct" to describe their absence from Ireland but, since he goes on to conclude that it is the same animal as the moose, it's clear he only means it in the regional sense. As possible reasons for its disappearance, he suggests an epidemic or overhunting by the early Irish. He specifically excludes the Deluge as a possible cause, saying that the horns would have been broken if they had been buffeted around by the flood waters, and, in any case, the flood was too long ago (over four thousand years by his reckoning) for the horns to have survived. As to how the animal once populated both Ireland and the New World, he says the only possibility is that the two must have once been connected.

Molyneux returned to the topic of giant bones in 1715. In July that year, Francis Nevile wrote a letter to the bishop of Clogher telling of the discovery of four enormous teeth during the construction of a mill in Northern Ireland. Nevile is of the opinion that, if they are those of an elephant, they must have been carried there by the Deluge, because no one in earlier ages could have had reason to bring an elephant to Ireland. Molyneux saw the letter and convinced the owner of the teeth to send them to him to examine. He had high-quality drawings made and wrote up the results of his examination for the society. He had no trouble identifying them as elephantine. To make his case he made liberal use of two published elephant autopsies, Mullen's and another by Patrick Blair of an elephant that had died in Dundee, Scotland, in 1706. Blair's autopsy was particularly detailed and filled two complete issues of the *Transactions* in 1710. Satisfied that he has established that they are elephant's teeth, Molyneux goes on to discuss how they got there. Once again, he rejects the Deluge and, referring to his earlier discussion of moose in Ireland, says that the "*Distribution* of the *Ocean* and *Dry-land* [of the earth] its *Islands*, *Continents* and *Shores*" must have once been different enough to allow elephants easy passage to Ireland. To bolster this argument, he refers to accounts of elephant teeth found in England and to Tentzel's account of the Tonna elephant. Finally, he mentions Ides's account of mammoth remains in Siberia. After Tentzel and Witsen, Molyneux is only the third writer to explicitly make the connection between elephant bones found in Europe and mammoth bones found in Siberia. The society published both letters and the drawings along with comments of their own (probably written by the secretary, Sir Edmund Halley, of comet fame). Sir Hans Sloane brought elephant's teeth from his collection for comparison, and several members made a field trip to Westminster to examine a complete elephant's skull held there. This satisfied everyone that the Irish teeth came from a young elephant, about half the size of the Westminster one.

Access to Sloane's collection, which was becoming recognized as the greatest natural history collection in Britain, was a useful asset in making the identification. Sloane was born in Ireland in 1660 and began collecting curiosities as boy. He studied medicine in London and France, collecting plants the whole time, and was well enough known that he was elected a fellow of the society soon after graduating. In 1687, he sailed to Jamaica as the personal physician to the new governor, the duke of Albemarle. The duke died after only fifteen months (through no fault of Sloane's). When Sloane returned with the duke's body, he had amassed a collection of eight hundred plants, mostly unknown in Britain, and scores of insects, artifacts, and other objects that caught his eye. He also brought back his own recipe for a healthful tonic made of cacao, sugar, and warm milk. In 1693, he became the secretary of the society and, therefore, the editor of the *Transactions*. In this capacity, he would have been the one who received Tentzel's pamphlet and the bones he sent. After marrying a rich widow and marketing his chocolate beverage, Sloane was able to acquire entire collections from other collectors, greatly increasing the size of his own. In 1722, John Bell passed through London following his diplomatic mission to China on behalf of Peter the Great. He brought with him a modest-sized but complete mammoth tusk, which he presented to Sloane. According to Sloane, Bell received the tusk from the wife of the governor of Siberia "in Lieu of a Reward for having cured her of a Distemper."

When Sir Isaac Newton died on March 20, 1727, Sloane was elected to take his place as the president of the Royal Society. Later that year, the *Transactions* published two papers he had written on the topic of buried ivory. These made up the most comprehensive survey on the topic up to that time and for a good time after. In the first of these, titled "An Account of *Elephants Teeth* and *Bones* found under Ground," Sloane describes the tusks in his own carefully cataloged collection. For each he gives a physical description, an account of its discovery, and other facts he thinks might be relevant. Following the paper are seven drawings of the tusks. His discussion of number 117 in his collection includes an essay on some historical discoveries, including Boccaccio's Sicilian giant and Tentzel's elephant. Number 1185 is the mammoth tusk Bell brought back from Siberia. It is in the context of describing Bell's tusk that he first brings up the word "mammoth." His discussion of mammoths takes up almost half of the paper. He liberally quotes or paraphrases Ludolf, Ides, Müller, and Tatishchev, along with the secretary on Bell's diplomatic mission, Lorenz Lange, and Cornelius de Bruyn, who had been with Peter the Great when he was shown the bones that he credited to Alexander's armies. In mentioning each author, he quotes their spelling or form of the word "mammoth." When he discusses the mysterious beast on his own he uses the form "Maman," which comes from Müller. His account of Bell acquiring the tusk would be the only published version of that discovery until Bell published his journals twenty-six years later.

The next issue of the *Transactions* contained the concluding part of Sloane's survey: "Of Fossile *Teeth* and *Bones* of *Elephants*. Part the Second." In this part, he leaves his collection and other bones that were available for him to personally examine and moves on to "several antient and modern Authors" who saw "Skeletons and Parts of Skeletons which are shewn up and down as undeniable Monuments of the Existence of Giants." Before diving into this topic, he digresses for a moment to compare the vertebrae of a whale and an elephant, showing how they should not be mistaken for a human's. With Mullen's and especially Blair's anatomies available, Sloane was confident he could recognize elephant bones when presented with them. What the educated world needed now was comparative anatomies of different species and categories of animals. Returning from his digression, he begins listing some giants' bones in literature, starting with the ancient authors, working his way through Boccacio's and Kircher's comments about them, and finally arriving at more recent times. He avoids the easy path of making a blanket declaration that everything is an elephant and looks for those accounts where the authors gave enough details, such as the size of a tooth or the dimensions of a skull, that he could identify elephantine features. Even without significant details, the presence of ivory always means "elephant" to Sloane. Among the ancients and church fathers, only St. Augustine's tooth meets his standards as being recognizably an elephant's. On arriving in more recent times, he feels confident in naming most reported giants as being elephant's remains.

He specifically mentions the Krems giant and Tentzel's elephant along with other recent discoveries in Italy, Germany, Poland, Greece, and Hungary. From the last country, Count Luigi Marsigli had brought back bones and ivory from several elephants (his collection later became the property of the Scientific Academy of Bologna, which he founded). Marsigli thought these were the remains of Roman war elephants. Sloane refers to the arguments of Tentzel—that Romans wouldn't have abandoned the ivory and that the undisturbed strata above them meant that they couldn't have been buried by humans—to decide against that interpretation.

In neither paper did Sloane have anything particularly new to add to what was already known about buried elephants and mammoths, namely that they were real organic remains. He added descriptions of his own collected pieces and Bell's account of how his tusk was found, but these were additional evidence of things already known. Nevertheless, the importance of the two papers is threefold. First was bringing all of this information together, which was an act that shouldn't be underestimated. Sloane's survey was, by far, the most complete done on the subject and would remain so for almost forty years. Second was decisively linking the mysteries of elephant bones in Europe and mammoth bones in Siberia together and calling it one unified question. What were these bones and how did they get to their far-flung locations? Third was the fact that this was all being done by Hans Sloane himself. At this point in his life, he had enormous credibility among Europe's intellectual elite. For twenty years, he had been the secretary for the society, a position that made corresponding with the greatest minds in Europe a duty. The extent and quality of his natural history collection was well known—later it would be the basis for the British Museum. When he said the elephant and mammoth questions were one and the same, his opinion had a weight that couldn't be dismissed. Typical of his present-the-facts-and-stand-back attitude was how he dealt with the question of how these elephants and mammoths came to buried where they were found. Using Tentzel's argument, he pointed out that human agency could not explain the best-documented discoveries, and he agreed that the Deluge was the best solution available. But that agreement didn't equal support. He allowed that other solutions might exist. After summarizing Tatishchev's paper, he added a plea for more information:

> It is to be hoped, that this Matter will one Time or other be set to a still clearer Light, particularly after the Order of his late Czarish Majesty was pleased to give to the Governor General of Siberia, to spare no Care nor Cost to find a whole Skeleton of this Animal.

Over in Russia, Peter the Great was impressed by the academies he met with during his two trips to the West. He added the creation of one, along with its own journal, to the long list of things he needed for a fully modernized country. Two problems stood in his way: a lack of educated Russians and the fact that most of the budget was going directly or indirectly to the military. For a time, the second of these slightly alleviated the first. The experts he hired to help with his military build-up educated a cohort of Russians in technical skills, mathematics, and basic literacy (though usually in Dutch or German). Following his first trip he corresponded with Leibniz about education, and although he didn't adopt his recommendations, it led him to experiment. Trying to build on the existing system of church schools was a failure due to the hostility to secular knowledge among church officials and the fact that none of the teachers knew much beyond how to train clergy. Once the war began he allowed POWs to teach some noble boys and even invited German missionaries to set up schools, but again the results were meager. After Poltava, he encouraged scores of Swedish officers to explore the east, but he must have known most of them would leave once the war was over. Sometime before his second trip to the West, knowing that the war was winding down, he began to seriously make plans for a university and academy after the war. Knowing that this was a few years in the future and being the impatient man he was, Peter hired a small number of experts to make more formal expeditions than the POWs were making. Most of these trips were planned by his doctor, Robert Erskine.

The model of these trips may have been that of Cornelius de Bruyn. With an introduction and funding from Witsen, he traveled down the Volga and Don in 1705 with the tsar and recorded his opinion of mammoth bones found near Voronezh before continuing on to study the Caspian Sea and explore Persia. In 1717, Peter sent Gottlob Schober to make the same trip and explicitly instructed him to observe natural history along the way. In 1715, Peter sent Lorenz Lange to China to hunt for the elusive opportunities for trade. He was pleased enough with Lange's work that he sent him on a second trip in 1719. It was on the trip that he and Bell examined the mammoth tusks that Bell would eventually present to Sloane. At Kiakta on the Chinese border they met Tulishen, the Kianxi emperor's envoy, who was heading west. It may have been during this meeting that Tulishen learned the word "mammoth" and made the connection that it and the *fyn-shu* were the same animal. During his second trip to the West in 1716, Peter stopped in Danzig and viewed the collection of Johann Philipp Breyne, a local doctor and fellow of the Royal Society. While there, Peter asked Breyne to recommend someone who could "undertake a voyage through Russia and make a description of everything remarkable." Breyne recommended his fellow physician, Daniel Messerschmidt. Messerschmidt had been taught by the same missionaries Peter had allowed to come to Russia to teach. From them he would have gained an idea of what to expect there. After some consideration, he accepted Peter's job offer and left for St. Petersburg in February 1718. After a year of planning, setting objectives, and gathering equipment, he left for Siberia in March 1719. In September, he arrived in Tobolsk and began interviewing the Swedish officers, including Philipp Tabbert von Strahlenberg.

Messerschmidt was heartbroken when Strahlenberg received his orders to return home. He wrote in his journal for May 13, 1722: "I separated myself with many tears from the pious, honest, hardworking, loyal Tabbert, my only friend and support. I am now left wholly abandoned, without society or aid. I will never forget my dear Tabbert." He promptly sank into a deep depression. His journals from this period include meditations on biblical verses related to the Apocalypse. But he didn't allow this to disrupt his work schedule. Everywhere he went, he made careful measurements of the latitude. He continued to collect samples. He was particularly thorough on the topic of birds, filling eighteen notebooks on this one topic. On January 16, 1724, while wintering in Irkutsk, he was presented with an almost perfect mammoth skull by Michael Wolochowicz. He prepared several precise and detailed drawings of the skull (front, rear, and profile views), a femur, a tooth, and a very mature tusk. These were of such high quality that they were still being used by foreign scientists at the beginning of the next century. In his thoroughness, he had Wolochowicz give sworn testimony regarding the circumstances of their discovery. Wolochowicz said that the bones had been found on the banks of the Indigirka River near the Arctic Ocean. Although a soldier, Wasile Erlow, was the actual discoverer, Wolochowicz went to the site to supervise the excavation. On the far bank of the river, he saw a piece of skin protruding from the earth.

On account of its size, he assumed it was from the same animal as the bones. He described the skin as, "very thick, and cover'd with long Hair, pretty thick set and brown, somewhat resembling Goats Hair: Which Skin I could not take for a Goat, but of the Behemoth; in as much as I could not appropriate it to any Animal that I knew." Messerschmidt sent the skull back to St. Petersburg with one of his periodic reports.

While still on the road, Messerschmidt sent the drawings of the head, the testimony of its discovery, a piece of ivory, and a tooth to Breyne. Among Messerschmidt's contacts, Breyne was an excellent choice to examine these new findings. He eagerly pursed the subject; he sought out and read existing accounts of discoveries and opinions about the mammoth and wrote a paper on the topic. From the materials that Messerschmidt sent him, Breyne was fully convinced that the mammoth was an ordinary elephant carried to Siberia by the Deluge. He drew no conclusions from the piece of skin covered by long hair, which Wolochowicz had not recovered but only detailed in his testimony notes. In 1728, he wrote a small paper on the teeth that was published in the journal of the scientific society in Danzig. This journal had no circulation beyond its members, and that was its final year of existence. Breyne's paper and Messerschmidt's drawings might have been forgotten at this point, but Breyne was determined not to let that happen, and he was a corresponding member of the Royal Society.

Strahlenberg left Messerschmidt because the Great Northern War had finally come to an end. Back in St. Petersburg, this meant that funds spent on the war were finally available for secular, nonmilitary projects. Against the recommendations of many of his correspondents, Peter chose to ignore the general lack of education in his country and approach the creation of his academy from the top down. The plan he approved was modeled on the French Académie des sciences. The scholars were to be employees of the crown working on projects assigned or approved by a royal appointee. The scholars were expected to give public lectures and privately tutor promising Russian students. He had several scientifically savvy advisers, including Tatishchev, recommend foreign scholars for recruitment. Johann Schumacher, his librarian and the curator of the Kunstkamera (his now extensive natural history collection), traveled west to buy books, develop contacts with other societies, and meet with the suggested scholars. The formal plan was approved in early 1724, and by the end of the following year, sixteen foreign professors had settled in St. Petersburg and taken up their duties. The director was Schumacher's superior Lavrentii Blumentrost. Peter donated the Kunstkamera and his personal library to the academy. In this first class were two who would add to the knowledge of the mammoth: the astronomer Joseph-Nicolas Delisle, the brother of the royal cartographer of France, and the young historian Gerhard Friedrich Müller. Two years later, Johann Georg Gmelin arrived to take up the chair of natural history. All three would later have opportunities to contribute to the understanding of the mammoth.

Peter did not live to see the academy in action. In November 1724, he fell ill. The illness developed into pneumonia, and he died the following January. Being Peter, he didn't let a little thing like dying slow him down. During his last weeks, he continued to plan new enterprises and dispatch orders. Admiral Fyodor Apraksin was called to his bedside to plan an expedition to the easternmost end of the empire to see if it was connected to North America and if any other European powers were in the region. By Christmas, they had completed the plan. In less than five weeks, they selected the personnel, ordered the supplies, and wrote all the necessary permissions. Vitus Bering and two other naval captains set off for the Pacific coast two days before the tsar's death on January 28, 1725. It took two years to reach the coast. Along the way, they gathered carpenters and blacksmiths to build their ships, they ordered supplies and sent them ahead, and they interviewed people knowledgeable about the land they would be crossing. At Yeniseiesk the met Messerschmidt heading west. The two stayed together for almost three weeks, exchanging notes and comparing maps. Messerschmidt was using a version of Witsen's map but, having made it most of the way to the coast, he would have had valuable new data to add to it. Bering was most likely using a map by Johann Homann, which included information from Strahlenberg's map. If Messerschmidt recognized the work of "dear Tabbert," no record has survived of it.

Messerschmidt was not carrying his newly acquired mammoth skull when he met Bering. His method of work was to finish each day by writing up his observations, carefully packing his samples, and categorizing each one according to his own system. About twice a year he would write up a report and send it along with the samples to Johann Blumentrost, the brother of the academy director. These became property of the academy after its establishment. His methods were so thorough that Shumacher brought some of his reports with him on his recruiting mission to show Westerners the kind of workmanship he was seeking. Messerschmidt sent the mammoth skull and his report on it to St. Petersburg two years before meeting Bering. When he reached the capital a year later, he was met with a cold reception. Peter's successors had little interest in his grand plans of exploration and had left the academy severely underfunded. The Blumentrost brothers ordered him to turn all of his journals and samples, including the skull, over to the Kunstkamera and limited his access to them. A committee that included Müller and Delisle made an official study of his materials and recataloged them according to the academy system. He was only permitted to keep a few of the duplicates that he had collected for himself. Had he not sent the mammoth teeth to Breyne, it is unlikely he would have been allowed to keep them, either. Many of his claims of compensation for expenses incurred while collecting samples were refused, too. Finally, bitterly, he packed up and headed back to Danzig. Along the way, he was shipwrecked and lost all his remaining belongings and samples.

That should have been the sad end of Daniel Messerschmidt, but he was called back for one brief encore. When Strahlenberg published his book on northeast Asia, he used considerable material that he had gained during his year with Messerschmidt, whom he remembered as a "worthy Friend, . . . [whose] Stay had been longer than mine, and who, as a Man of Letters, might probably have been fitter for this Work than myself." This brought him to the attention of Tatishchev, who pushed the academy to bring him back to St. Petersburg in 1731 to edit his journals and notes and organize his collections so that they might be of use to other scholars. Almost immediately, they were put to use. Bering had completed his journey in February 1730 and met with almost as a cold a reception as Messerschmidt had received. The admiralty responded to his report with a yawn and filed it away his plans for follow-up missions with no comment. Things changed the following year. Anna, the new empress, had among her inner circle several supporters of the academy and Peter's program of exploring the east. The academy was once again fully funded, Bering's plan was pulled out of the admiralty files, and plans were made to enact it. In preparation, several of the principals of the expedition availed themselves of Messerschmidt's expertise and notes.

The Second Kamchatka Expedition, or Great Northern Expedition, grew to be one of the largest, most comprehensive, and most expensive scientific and exploratory missions in history. By the time they set out in 1733, it included three components. The first was a series of small expeditions with the goal of mapping the entire Arctic coast of the empire. The second was to proceed to the Pacific coast to build ships. They then split into two groups with one attempting to find a trade route to Japan and the other sailing due east to find the North American coast. The latter would be under the personal command of Bering. The third component was a group from the academy, led by Müller and Gmelin, who were to conduct scientific and historical research in Siberia and support Bering as needed. The academic contingent included more than a dozen students, artists, and technicians, one of whom was Joseph-Nicolas Delisle's ne'er-do-well half brother Louis de l'Isle de la Croyère as an astronomer-cartographer. The academy contingent was to leave Moscow in late summer, catch up with Bering in Tobolsk, and continue with him to the Pacific. When they left on August 8, 1733, they had with them forty wagonloads of supplies, including nine wagons of scientific instruments, a small library, writing and painting supplies, and several kegs of Gmelin's favorite German wine (for medicinal purposes), all protected by fourteen soldiers and a drummer.

The first year on the road for the academics was productive and almost pleasant. They leisurely made their way to Tobolsk. Gmelin collected and described plants, while Müller combed through local archives for historical records. De la Croyère attempted to make astronomical observations, but was in far over his head. Their stay in Tobolsk overlapped Bering's only for a short while. When they left, rather than following him to directly to Irkutsk, they followed the Tom River to its headwaters. From here, the trip became less pleasant as they encountered uncooperative officials, impassable landscapes, and Siberian mosquitoes. Almost two years passed before they caught up with Bering in Yakutsk. Bering, weighed down by administrative and logistical matters, had yet to advance to the coast for his own voyage to North America. He and the academics took a strong dislike to each other.

In order to avoid the commander, Müller and Gmelin began making short trips into the countryside. It was either during one of these side trips or on the road home that they encountered the mammoth (neither gives a date).

Müller, for his part, gives a description of the trade in mammoth ivory, which he considers one of the important commodities of Siberia. In his opinion, mammoths cannot be anything other than real elephants brought there by some great cataclysm in the earth. He does not claim that the cataclysm had to be the Deluge in particular. He describes how and where they are found and the products manufactured from them. Gmelin goes into much more detail than his colleague. In his account of the expedition, published in 1752, he dedicates twenty-five pages to the topic, making it one of the longest expositions on the mammoth up to that date. His conclusions are nothing new. The belief that the mammoth was an ordinary elephant carried to Siberia by the Deluge or some other catastrophe was approaching accepted wisdom by the time he wrote. The value of his account is that he makes some interesting connections and adds descriptions of of previously unknown discoveries. He describes two different discoveries that were reported to Yakutsk soon after Peter issued his orders that any good bones be sent to the Kunstkamera. For each, he describes the places they were found, riverbanks above the Arctic, and gives the names of the discoverers. He makes the connection between Siberian mammoths and elephant bones found in Central Europe and Dauphine, where the Theutobochus bones were found. He gives a good deal of space to the trade in walrus and narwhal ivory and how the three are combined and confused. His unique contribution to the mammoth question is that he describes the skulls of two other unknown large animals dug up in Siberia. Both skulls he says resembled large oxen. The first was described for him by the governor of Yakutsk. In 1722, he told Gmelin, a hunter named Spiridon Portniagin had returned from a trip to the Arctic coast with a good ivory tusk. Near the spot where he found the tusk, he found another skull that resembled a bull's head, but with the horns on the nose. When the governor asked him to return to the spot and recover the skull, Portniagin begged off, pleading old age and an eye infection. At the same time, the governor was able to give Gmelin another skull that really did resemble an ox. Gmelin shipped the skull back to St. Petersburg. As for the skull with its horns on its nose?

In another place, Gmelin mentioned his distrust of the wild tales that the Siberians told about the mammoth. He must have relegated this report to the same category. He didn't pursue the topic any further. Taken together, his notes hint at a Siberia much stranger than had he and other Europeans and city-dwelling Russians had suspected.

During their time in Yakutsk, Müller and Gmelin decided that they really didn't want to go to the coast after all. Aside from their personal dislike for Bering, both had suffered serious bouts of ill health and wondered if they would even get out of Siberia alive. To avoid being completely negligent in their duties, they assigned de la Croyère, who had long since proven likable but incompetent, and Stephan Krasheninnikov, their most promising student, to go in their places. At Yeniseisk, on their return journey, they met Georg Wilhelm Steller, an adjunct academician who had been sent out to become Gmelin's assistant in response to a request he had made three years earlier. The two senior academicians decided to send the young man to join Bering. Steller was a German physician who arrived in St. Petersburg a year after the academic contingent left, with little money in his pocket and no more coherent of a plan than seeking adventure. Through luck or plan, he soon met Theophan Prokopovich, the archbishop of St. Petersburg and a confidant of the late tsar, who had taken part in the discussions that led to the creation of the academy, and Joseph Amman, a fellow German and keeper of the academy's botanical garden. The two older men were so impressed by Steller that they sponsored him for membership in the academy. At about this time he also sought out and met Messerschmidt, who was living in bitter retirement. The two had both studied medicine under Friedrich Hoffman, but it was Messerschmidt's stories of Siberia that held them together. During the last few months of Messerschmidt's life, the two spent many evenings together talking about science and adventure in the far east. In one of these conversations, Steller learned about the mammoth. As a member of the academy Steller would have had access to the bones in the collections.

From the moment he heard about the Great Northern Expedition, Steller desperately wanted to be a part of it, but he was a year too late to join. Amman had him help with his botanical projects, one of which was organizing Messerschmidt's samples, giving him a chance to observe firsthand the young doctor's vast knowledge of natural history. When Gmelin's request for an assistant arrived, he was more than happy to recommend Steller for the job. After more than a year of bureaucratic delay, Steller's contingent left the capital at the beginning of 1738. He chose a route that would take him to Yeniseisk by different rivers than Gmelin had so that he would be able to collect a different set of plants. After they met, Gmelin and Müller wasted no time in sending him on to join Bering while they continued west. Steller's new duties and adventures were not enough to make him forget mammoths. After joining Bering, sailing to Alaska, being shipwrecked on a barren island, and barely surviving to return to Siberia, he immediately planned a solo expedition to the Kolyma River to hunt for mammoths. In the introduction to a book on beasts of the sea, he wrote, "My zeal is fired by those mammoth skeletons and the slight accounts of them." He only made it halfway to his goal. Two years later, exhausted by ill health and increasingly bitter fights with local officials, he began the journey back to St. Petersburg. He died on the way. Leonhard Stejneger, Steller's biographer, believes he was thinking about mammoths right up to the end. In one of his last communiques with the academy, Steller mentioned an unspecified plan that he wanted to propose after his return. Stejneger was sure this was to be another trip to the Arctic coast to look for mammoth remains.

Finally, out of this first generation of academicians to come of age in the latter half of the seventeenth century and the early decades of the eighteenth, we have Joseph-Nicolas Delisle. When Delisle sent his semi-incompetent half-brother to Siberia with Gmelin and Müller, he wasn't trying to avoid the rough life of exploring. While the Great Northern Expedition was exploring Siberia, he went on several expeditions of his own. In early 1740, Delisle made a trip to western Siberia to observe a transit of Mercury across the sun. After six weeks of travel, it was cloudy on the day of the transit and he was unable to make any observations. Like most members of the Russian Imperial Academy, he wasn't one to waste a good trip. He made observations and collected data about many topics outside his field. On his return trip, he heard that a monastery in Tobolsk had some mammoth bones. The brothers showed him the bones and told him stories about some discoveries, including one by a famous bear-wrestling

Cossack. After that, Delisle made a point of adding some bones to his personal collection. Soon after his arrival in Russia, officials began to suspect that he was copying geographical information and forwarding it to France. An official investigation was launched, and he was carefully watched. In time, the investigators decided there was no evidence of wrongdoing and reduced their surveillance. He remained in Russia for twenty-one years and retired with a full pension. Their suspicion that he was feeding information to France turned out to be true, but he was a better spy than they were spy catchers. When he returned to France in 1747, he brought with him more geographic information and the femur of a mammoth.

It was September 1735 before Breyne gathered his mammoth data and sent it to Sloane in London. No doubt his impetus for doing so was learning of Messerschmidt's death earlier that year. Breyne's package included an English translation of his own paper from seven years earlier, the affidavit of discovery by Michael Wolochowicz, Messerschmidt's drawings of the mammoth's skull with detailed measurements, and a short letter summarizing his conclusions. Sloane presented Breyne's materials to the Royal Society in early 1737 and had them published in the *Transactions* for that year. Breyne's paper was brief and summed up what was already known. He made references to some of the Swedish prisoners' accounts and stated that he was convinced that the bones were evidence of the Deluge. Wolochowicz's affidavit was of historical, not scientific, interest. The true value in his package and Sloane's publication of it was in the drawings and measurements. The society's printers produced superb plates from them. These provided savants all over the continent and in the colonies with clear images to compare with their own discoveries. They served the same purpose for mammoths that Mullen's drawings had served for elephants sixty years earlier. Appropriately, Breyne had used Mullen's plates to make his identification.

With that, the mystery was solved to most people's satisfaction. The giant's bones of Europe and the mammoth's bones of Siberia were actually from the same animal, and that animal was an elephant, a beast known to the church fathers and the greatest writers of antiquity. And these bones had been brought to their disparate locations all around the world due to a cataclysmic event, whether it was the Deluge or another catastrophe.

There was no reason to reach for fantastic explanations like sea monsters or carriage-sized moles. The men of the academies and societies had now seen, autopsied, and carefully described enough real elephants and compared their bones to those of mammoths to be confident that they were one and the same. Any differences could be explained as the normal range of variation to found in any species, such as dogs, hogs, or cattle. And then word began to come in from the Americas about more large bones that didn't quite fit with this happy and neat solution.

CHAPTER 6

THE AMERICAN COUSIN

In the fall of 1519, Hernán Cortés, six hundred soldiers, and fifteen horses arrived in the territory of the Tlaxcala, on the eastern edge of the central valley of Mexico. Tlaxcala was one of the last Nahua states to remain free of Aztec rule. For decades, they had been locked in the horrific Flower Wars with their larger neighbor. The purpose of these wars was not conquest or settling geopolitical conflicts but to provide prisoners to sacrifice to the gods to stave off famines caused by overpopulation. The first impulse of the Tlaxcala was to destroy this new alien army, and they had an opportunity to do just that, but divisions among their leaders led them to recall their army and make peace with the Spanish. They realized that the newcomers might be able to break the horrible stalemate of the Flower Wars in their favor. They invited Cortés into their city to rest and negotiated an alliance against the Aztec capital of Tenochtitlán. While the Spanish rested in Tlaxcala, preparing for a new campaign against the

Aztecs, the Tlaxcala leaders made every effort to curry favor and impress these strangers. The Spanish were fed and entertained. The leading houses allowed their daughters to be baptized. At some point during the three weeks the Spanish stayed in Tlaxcala, a group of Spaniards began to question their hosts about their history. Bernal Díaz del Castillo, a soldier under Cortés's command, who wrote a history of the campaign, recorded their answer.

> They said that their ancestors had told them, that in times past there had lived among them men and women of giant size with huge bones, and because they were very bad people of evil manners that they had fought with them and killed them, and those of them who remained died off. So that we could see how huge and tall these people had been they brought us a leg bone of one of them which was very thick and the height of a man of ordinary stature, and that was the bone from the hip to the knee. I measured myself against it and it was as tall as I am although I am of fair size.

The Spanish helped themselves to the bone and sent it to King Charles I on the first treasure ship out of Veracruz. The bones of both Columbian mammoths and American mastodons have been excavated in that part of Mexico, but a bone as long as the one Díaz described probably came from an earlier mastodon species such as *Rynochotherium tlascalae*, which has been found in that region and better fits the dimensions that Díaz mentions. In researching her book *Fossil Legends of the First Americans*, Adrienne Mayor went searching for this femur. Although officials at the national museums in Spain couldn't identify that specific bone, they didn't exactly rule out its being there. The records for those years are just too sparse to be sure. It very well might be that the Tlaxcala femur is sitting, unlabeled, in a warehouse somewhere in Madrid.

Díaz wasn't the only Spaniard to report on the presence of large bones and legends of ancient giants. Cortés himself had a collection of giant bones at his estate all collected in the New World. Later travelers José de Acosta, Antonio Herrera y Tordesillas, and Joseph Torrubia were also shown giant teeth and bones by the Tlaxcala and recorded the legends of giants that the native Mexicans believed. Nor was the valley of Mexico the only place where such bones were reported by the Spanish conquerors. In South America, during the chaos following the destruction of the Inca empire, two different writers, Pedro Cieza de León, a soldier, and Agustín de Zárate, a clerk, described bones on the on the Santa Elena Peninsula near the site of modern Quito, Ecuador. The legends of the Manta natives who lived there were different than those of the Tlaxcala. In their mythology, the bones were of insatiable giants who invaded the country, murdered the people, and ate all the food. The people were defeated in every attempt to fight the giants. Finally, the gods heard their prayers and destroyed the giants with lightning.

Neither Cieza nor Zárate was able to go to the peninsula to see the bones. Cieza heard enough from Spaniards who had seen giants' bones in other parts of the Americas to accept that the story must be true, though he though they were probably exaggerated. Zárate wrote that the story seemed too fantastic to believe until he heard of another Spaniard who had made the trip to the peninsula specifically to see these bones. This man, Captain Juan de Olmos, the lieutenant governor of Puerto Viejo, was able to find and subsequently excavate these bones in 1543. What Olmos did was quite advanced for his time. He could easily have ordered the natives to bring him a few bones. Instead, he went to the place where the bones had been reported and examined them in situ before beginning careful extraction. He recovered the complete bones of this being and tried to reconstruct what it might have been based on the knowledge and worldview that he had. The bones that Olmos examined were most likely those of *Cuvieronius hyodon*, a type of gomophthere (relatives of the mastodon) commonly found in the area. The only specific detail he describes are the teeth, which don't match any other animal since discovered in the region.

Even though the Tlaxcala and Manta came to the same conclusions as sixteenth-century Europeans when presented with proboscidean skeletons—that is, that they were the bones of giants—they were, in many ways, far ahead of the Europeans in the practice of paleontology. The Native Americans never questioned the organic nature of the bones. This was true of petrified bones, such as dinosaur fossils, as well as nonpetrified bones, such as mammoths and mastodons. They had no equivalent theory to the European *vis plasica* holding back their understanding. Native Americans had a second conceptual advantage over Europeans. They had no trouble with the idea of extinction in their worldview, unencumbered by the Christian version of divine creation and the divine chain of being. Whether the giants were killed by Tlaxcala warriors or smited by the gods, they were a race of beings now gone from the earth. As we've seen, it was not until the middle of the eighteenth century that *vis plastica* was fully defeated. And discomfort with the idea of extinction among Europeans would persist well until the beginning of the following century.

The books containing the Spanish descriptions of giant's bones in the New World were not obscure. In particular, Father Acosta's *Natural and Moral History of the Indies* was broadly known. In the thirty-five years after its first appearance in Latin in 1590, it went through over a dozen editions and was translated into Spanish, English, French, Dutch, and German. Herrera y Tordesillas's history, known as the *Décadas*, which repeated Zárate's story, was translated into the same languages. Other authors writing about giants repeated the Spanish stories. And at least one writer made the connection between giants and elephants. That person was our friend Nicolaas Witsen, who, in his two books on shipbuilding, mentions "Elephant Teeth that can be got in Mexico from the deep abyss of the earth."

The first published account of giant bones in Anglo-America appeared in the *Boston News-Letter* on July 30, 1705. The bone was a giant tooth dug out of a hillside by a farmer near Claverack, in the Hudson valley, twenty miles south of Albany. The article described it as "a great prodigious Tooth . . . supposed by the shape of it to be one of the far great Teeth of a man; it weighs four pound and three quarters, the top of it is as sound and white as a Tooth can be, but the root is much decayed, yet one of the fangs of it hold half a pint of Liquor." Peter Van Bruggen bought the tooth for half that much liquor and presented it Lord Cornbury, the governor of New York. Cornbury, in turn, packed up the tooth and sent it to the Royal Society with the request that they give it to Gresham College after they were done examining it. Cornbury was not a popular man in the colony. A few years later he would be indicted for embezzlement, graft, and general obnoxiousness, including walking the streets of New York in a dress, an act which was believed to be mocking Queen Anne. From our perspective, he did have one redeeming characteristic: curiosity. After receiving the tooth—and possibly emptying it of any remaining liquor—he ordered the Recorder of the Assembly, Johannis Abeel, to investigate the place where the tooth was found, and later he traveled there to look at it himself. Abeel reported finding a skeleton thirty feet long, but said that the bones had crumbled when they tried to remove them from the ground. The *Boston News-Letter* also mentions bones that could not be recovered, including "a Thigh-bone [that] was 17 Foot long."

The following summer, more teeth and bones were found on the Hudson. These were brought to the attention of Cotton Mather, of Salem witch trial fame, and, through him, to the Royal Society. In a letter to Mather, Massachusetts governor Joseph Dudley described the circumstances of the discovery. "[T]wo honest dutchmen" brought him some teeth and bone fragments. These, they said, were found near Claverack in a pocket of earth seventy feet long, of a different "colour and substance" than the surrounding soil. This discolored earth, they assumed, was the remains of the body of the creature that had previously owned the bones and teeth. One tooth was almost perfect. Dudley said it was of the same configuration as Cornbury's tooth and that he and all the doctors of the town agreed that it exactly matched a human molar. It had to be from a human giant "for whom the flood only could prepare a funeral; and without doubt he waded as long as he could to keep his head above the clouds, but must at length be confounded with all other creatures, and the new sediment after the flood gave him the depth we now find." Dudley added that it was impossible that the tooth could have come from a whale or elephant. Claverack was too far from the sea for whales, and the tooth was the wrong shape for an elephant. The poet William Taylor heard of the same discovery and wrote that Indians from all around came to say "I told you so" to the rude Europeans who had discounted their ancestor's tales of giants.

Dudley's letter came a perfect time for Mather. For years, his political and social influence had been declining. Under Queen Anne, High Anglicanism was on the rise and Puritanism in decline. Mather's difficult personality also drove any remaining supporters away from him. His wife had died four years earlier, and he was in ill health. He was increasingly worried that he might not live to complete what he saw as his most important work, *Biblia Americana*, a gloss on different verses of the Bible in the light of modern scientific thought. He saw the giant bones and teeth of Claverack as the proof he needed for Genesis 6:4: "There were giants in the earth in those days." The gloss on this verse is one that he would be particularly proud of. In November 1712, Mather sent thirteen letters to the Royal Society over a twelve-day period. The mail being what it was, these letters almost certainly traveled together on the same ship. The letters appear to have been written over several years, despite being mailed within two weeks. One can only speculate as to why he didn't package them into a single bundle.

For Mather, and for us, the most important letter is the first. This letter is addressed to John Woodward, a member of the society who, in the 1690s, had written a rebuttal to Burnet's theory of the flood titled *Essay toward a Natural History of the Earth*. Woodward's countertheory challenged Burnet's idea of a pristine smooth earth before the Deluge, pointing out that mountains are mentioned in Genesis before the Flood, and took issue with the implication of great age for the earth in Steno's explanation of stratification. In his theory, the violence of the flood destroyed the old mountains, the sediment-filled waters next laid down all the visible geological strata, and the receding waters carved valleys into the still-soft sediment, creating the mountains of historical times. Woodward rejected the idea of *vis plasica* and embraced the organic nature of fossils. Within the violent, muddy waters of the Deluge, bones and shells survived and were eventually deposited in layers according to their various specific gravities. This is essentially the flood geology theory embraced by American Creationists today. Although the theory already had its critics, Mather felt that Woodward's theory best explained the strata visible in the earth and the placement of fossils in those strata. He believed his gloss of the giants fit in well with Woodward's theory and hoped to gain the patronage of his better-known peer. In this letter, he rather disingenuously presents his *Biblia Americana* as the work of a brilliant young colleague to whom he wishes to draw the attention of the learned men of Europe. After praising this anonymous author, he adds his gloss of Genesis 6:4 as a proof of his genius. After citing the usual ancient and medieval sources on giants, Mather describes the Claverack teeth. For Cornbury's tooth, he quotes, with credit, the newspaper account of the discovery. For the second tooth, he plagiarizes Dudley's letter, including the conclusion that the bones were found too far from the sea to have been from a whale and were of the wrong shape to have been from an elephant. In a long marginal note, he points out that such teeth and bones have been found by "Americans in the Southern Regions" and credits Zárate, Acosta, and Cieza as his sources. He finishes with a bit of patriotic pride: "But at last we dig them up in [our] Northern Regions too." He, too, describes the Indians coming to look at the bones and cite them as proof of their legends, but his reason for mentioning them is nothing more than a hook to launch into a short rant on how disgusting he finds the language of the "Salvages."

The Royal Society was not especially impressed with Mather's letters. Two years after receiving them, Edmund Halley, Sloane's replacement as secretary, published a short summary of them, giving each one a single paragraph, in some cases only a single sentence. As to the teeth, Halley repeats the physical descriptions of them, ending with the comment, "It were to be wish'd the Writer had given an Exact Figure of the Teeth and Bones." From the descriptions given, the teeth found at Claverack and in other parts of the Hudson valley came from mastodons. Like the Deinothere teeth that were mistaken for Theutobochus, mastodons' teeth have a hard enamel surface and cusped shape that closer resembles a human molar than an elephant's grinder. The Hudson valley is a very rich area for mastodon bones, and the most important find of the century would be made at Newburg, sixty miles downriver from Calverack. That would happen at the end of the century, long after Mather had died. For the time being, the Hudson valley teeth remained a local curiosity and nothing more.

The first clear mention of a mammoth in the Americas appears in 1743, though the discovery must have happened some years earlier. In that year, Mark Catesby published the second volume of his *Natural History of Carolina, Florida, and the Bahama Islands*. Like Mather, Catesby was a firm believer that the New World showed convincing proof of the truth of the Deluge. He gave several examples of sharks' teeth, shells, and the bones of other sea animals found deep in the earth as far as sixty miles from the sea. His last example can only be a Columbian mammoth: "At a place in

Carolina called *Stono*, was dug out of the Earth three or four Teeth of a large Animal, which, by the concurring Opinion of all the *Negroes*, native *Africans*, that saw them, were the Grinders of an Elephant, and in my Opinion they could be no other; I having seen some of the like that are brought from *Africa*." These unnamed African slaves, at some date before 1743 and probably before 1739, when a slave revolt threw the district into chaos, were the first non-Indians to make a relatively correct identification of a fossil vertebrate in the New World. The closest any European competitors came was a possible tie. In 2014, after hearing the story of the Stono teeth, eight-year-old Olivia McConnell successfully lobbied the legislature of South Carolina to make the Columbian mammoth the state fossil based on this find.

The middle years of the eighteenth century saw Britain and France in a struggle for North America. The two countries had pursued different strategies for colonizing the continent. The British settled the east coast and moved westward as the French settled the St. Lawrence and Mississippi valleys and moved outward. Both strategies required managing a network of constantly shifting alliances with Native American nations who, naturally, had their own interests to protect and worked on managing the French and British. By the 1720s, the French from Quebec had portaged from Lake Michigan into the Illinois River valley and established trading posts there, but expansion from the Caribbean up the Mississippi had stalled north of Natchez. The French hoped to monopolize trade coming out of the Mississippi valley, but when they moved north of Baton Rouge, they found that the British had already established a trade relationship with the Natchez and Chickasaw nations using an overland route from South Carolina. Supported by the British, these nations blocked French trade farther upriver. After multiple attempts to lure them away from the British failed, the French decided to use force to open the river. By 1731, the French had destroyed or scattered the Natchez. The Chickasaw proved to be a more difficult problem. In 1735, Jean-Baptiste Le Moyne de Bienville, the governor of Louisiana and founder of New Orleans, gave up on negotiation and decided that, once again, war was the only way to deal with his intransigent neighbors. A great campaign was planned for the spring of the following year. The plan was a simple one: one army would come down the Mississippi from the Illinois country while a second would come from New Orleans through modern Alabama, and they would crush the Chickasaw between them. The campaign was a miserable failure. The two armies failed to coordinate their actions, arriving and attacking days apart, and the Chickasaw defeated them one at a time, inflicting a great number of casualties on the French. Bienville returned to New Orleans to plan a second campaign.

Bienville began his second campaign in the summer of 1739. This time, the New Orleans force was reinforced with cannon, mortars, grenades, thousands of pounds of powder and shot, and five hundred troops fresh from France. A second force, under Pierre-Joseph Céloron de Blainville, was to come down from the Illinois country while a third, commanded by Bienville's nephew, Charles Le Moyne de Longueuil, was to come down from Quebec. Longueuil's force was made up of 123 French and Canadian and 319 fighters from allied Indian nations (186 de Sault Iroquois, 51 Two Mountains Iroquois, 32 Algonquin and Nipissing, and 50 Abenaki). Their planned route was to be almost entirely by water, up the St. Lawrence, across Lake Ontario, a portage around Niagara Falls, across Lake Erie to a place where they could portage into the headwaters of the Ohio River, and down that river to the Mississippi. This route allowed the expedition to perform a second service to the authorities in Montreal and New Orleans. The Ohio River was barely known to the French. By following this shorter route, Longueuil was able to assess whether it was superior to the established route through the Great Lakes and over the Chicago portage into the Illinois River. For this purpose, he was provided with a young surveyor, Joseph-Gaspard Chaussegros de Lery, the eighteen-year-old son of the chief engineer of New France. Longueuil's army left Montreal in several detachments over the last two weeks of June 1739.

A few miles below the site where Cincinnati would later be built, the army camped at the mouth of a stream on the southern bank of the river. De Lery noted on his map of the Ohio River that Longueuil made a formal showing of the arms of the king, claiming the land. De Lery called their camp "[The] place where the bones of many elephants were found." One of the officers, Major François-Marie Le Marchand de Lignery, wrote in his journal how these bones were found. A group of Indians went hunting for fresh meat to reprovision the army. Somewhat later, a few of them returned, bringing with them a giant femur and tusks that the French officers identified as coming from an elephant. The hunters said that there were three skeletons of this animal in a salt lick not far from the camp. If the hunters were local Shawnee, who had been recruited to replace various deserters, they would have been familiar with the country and already known that the bones were there. They probably made a special trip to collect them and show the Frenchmen in order to impress them. Or, maybe they brought them back to show the other Indians as proof of their story. In either case, Longueuil recognized the importance of the bones and had them added to his baggage. The French officers made the five-mile trip to examine the site and collected other bones as souvenirs. Longueuil added three teeth to his baggage. The bones from the Ohio were not a secret. Longueuil's officers told Bienville's officers about them and showed off their own souvenirs. One conversation that we know for sure happened was between Lignery and Bienville's secretary, André Fabry de la Bruyère. Lignery referred to his journal while telling the story. The journal has since been lost, leaving Fabry's memory of his conversation with Lignery the only account of the discovery that has survived.

Militarily, however, the campaign against the Chickasaw was no more successful than the previous one. The cannons got stuck in the mud, draft animals died, the French soldiers got sick, draftees deserted, and the Iroquois made a separate peace with the Chickasaw after exchanging gifts of cheese and pottery. Rather than meeting the French in battle, the Chickasaw wore them down by refusing to engage them. In the late spring of 1740, the French had suffered as many losses to disease as they had to battle in 1736—more than five hundred men. At this point, Bienville called off the campaign and released the armies—at least, that part that hadn't already deserted—to go home. Longueuil and some of his officers joined his uncle and escorted the sick soldiers downriver to New Orleans. As long as they were in North America, the French never did manage to defeat the Chickasaw.

Longueuil returned to France in the fall with the bones and teeth. He donated them to the Cabinet du Roi (the museum of the king), perhaps to salvage some goodwill after his military defeat. At the time of their donation, the bones failed to attract much attention. Academy member Henri-Louis Duhamel du Monceau was the only French scientist who was aware of the donation at the time. It is from him that we know of the three teeth that were donated along with the tusk and femur. It's curious that Georges-Louis Leclerc, Comte de Buffon, the new director of the Jardin des Plantes, which included the Cabinet du Roi, only learned about the donation when Fabry spent the winter in Paris in 1747–1748. Fabry wrote a short note describing the discovery as he heard it from Lignery and orally told Buffon about legends along the Mississippi, calling the skeletons the father of bison (*le pere aux beufs* literally translates as "the father of oxen").

Buffon's ignorance of the discovery is doubly curious, considering that word of the discovery had been published twice as notes on important maps. Philippe Mandeville, a member of the Louisiana contingent, used de Lery's notes to draw the first map. On it he made the notation "Place where the bones of many elephants were found by the army from Canada commanded by the Baron de Longuille [*sic*], and where he had the Arms of the King set up in 1739." When Mandeville's manuscript map was used by Jacques Nicolas Bellin in his new map of North America published in 1744, the engraver made a mistake and marked the date as 1729. In 1756, in a letter published in 1763, Jean-Bernard Bossu wrote that the discovery was made in 1735 during Bienville's earlier unsuccessful campaign against the Chickasaws and inflated the number of skeletons from three to seven. Confusion about the year of the discovery has persisted ever since. Longueuil's visit was definitely in 1739, and it is the earliest recorded visit by a European to the site, now called Big Bone Lick State Park in Kentucky. But it wasn't the last. In 1755, the British printer Lewis Evans published a "Map of the Middle British Colonies in America." Along the Ohio River was a small notation "Elephant Bones found here." Evans could have received this information either from Bellin's map or from British traders who had begun to move into the Ohio valley by then.

Through the 1740s, the bones of the Ohio, like the teeth of the Hudson, remained more of a local curiosity then objects of serious study. This began to change in the early 1750s. By then, the colonies were beginning to develop their own intellectual institutions. Along with universities, English

America got its own scientific society in 1743 when Ben Franklin and a group of like-minded men founded the American Philosophical Society. In 1752, the Swiss geologist Jean-Étienne Guettard presented a paper before the French Académie des sciences. The topic was a comparison of the geology of Switzerland and North America. Guettard had an interest in fossils and included a short section on the fossils of North America. Guettard's paper was published in the journal of the academy in 1756. In what appears to be addition to the paper at the time of publication, the paper includes two plates of a large tooth and a small crinoid fossil. The tooth is easily identifiable as coming from a mastodon. These are the first published images of mastodon teeth. In fact, they are the oldest surviving drawings of them in any medium. Guettard's images show a long tooth with ten conical cusps (one of them broken off) and long roots. The cusps are separated by deep valleys. The upper part of the tooth appears to be made of enamel like a human tooth. As was often the case with etchings of the time, the images are reversed. Of the plates, he writes: "I should have so much desired to compare a large fossil tooth that is from the place that is marked on maps of Canada as the canton where elephant bones are found. What animal is it? And does it resemble fossil teeth of this size that we have found in different parts of Europe? I present this figure; the research we do on it later should shed some light on the subject." He goes on to say that Jean-François Gautier, a prominent Canadian naturalist, had sent him a note commenting on his draft. Gautier wrote, "All those who have been to this place, who have seen the skeletons or bones of these animals, relate that the skeletons are almost complete. We do not assume that they include the teeth, because these are the only parts that we can easily carry away. The other bones are too large and too heavy." Gautier added that he will have Father Bonnecamp, a Jesuit of impeccable scientific credentials, make drawings of the skeletons during his next trip down the Ohio. Bonnecamp had traveled down the Ohio in 1749, though not as far as the Big Bone Lick. He never made another expedition, and Guettard's works make no mention of him examining any other "elephants'" teeth.

There is some mystery that remains about this tooth. Though most modern reproductions of the plates identify the tooth as one of Longueuil's, it does not match any of the descriptions or drawings published at the time of the teeth Longueuil brought back. In 2002, Pascal Tassy went through the fossil collection at the museum trying to identify Longueuil's bones. In the 260 years since their donation, the bones were cataloged three times and given different numbers each time. Old numbers rubbed away, tags fell off, and the collections were moved several times. But Tassy was successful and located all three teeth with traces of their original identification numbers visible under a black light. He also located the tooth in Guettard's illustration. After some detective work, he was able to match the tooth to a number in an 1861 inventory with the notation "Collection Drée." From here, he was able to find an illustration of the tooth and a reference to Drée in Georges Cuvier's *Recherches sur les Ossemens fossiles* (1806). This makes the person who donated the tooth to the museum, Étienne, Marquis de Drée, a prominent politician who aided Cuvier's researches by allowing him access to his collections. Drée, however, was not born until 1760. The tooth had to have had at least one other owner before Drée added it to his collection. The simplest solution is that Gautier sent it to Guettard sometime before 1756 and that he sold or gave it to Drée sometime before his death in 1786, but a definitive answer remains elusive.

Meanwhile, back at the place where these teeth were discovered, British Americans had no interest in respecting the French crown's claim to the Ohio country. Within a few years, possibly months, of Longueuil's passage, British traders from Virginia and Pennsylvania began working the valley. In the 1750s, hostilities broke out between French and British in the valley. This soon escalated into real war, with both sides sending military expeditions enforce their claims. In 1756, the Seven Years War broke out in Europe with Britain and France on opposite sides. Today, in the United States, the colonial front in the war is called the French and Indian War. After some initial victories, the French were driven out of the Ohio valley. The American phase of the war was essentially over by the end of 1760, when the British conquered Quebec. In Europe, the war continued for another three years, ending in the defeat of France and its allies. According to the terms of the Treaty of Paris, France gave up all of its territories on the mainland of North America. During these years, there was no opportunity for the French to collect new bones for study. This was not the case for the British.

In the years before Guettard's paper was published and hostilities broke out in North America, French informants were not the only people passing around stories of bones and actual bones from the country surrounding the Ohio. British merchants found more massive bones and told the educated men of Philadelphia and New York about them. Unlike Gauthier's vague references, specific dated descriptions of discoveries were shared in Anglo America and made their way into print in the 1760s. Soon after Longueuil's army returned to Quebec, a merchant named Robert Smith settled in the valley of the Great Miami River, a tributary of the Ohio, which debouches a few miles above the same salt lick. On the opposite side of the river, Smith discovered the salt lick and its bones in 1744. He later told the surveyor Christopher Gist that he saw three skeletons there, possibly the same ones Longueuil's hunters reported five years earlier. Smith also told Gist about the tusks, saying they were five feet long and very heavy. He tried to carry one away, but ended up hiding it so the French wouldn't take it. He failed to say if he ever retrieved it. Gist, who recorded Smith's story, was hired in 1751 to survey the river down to the Falls of the Ohio where present day Louisville, Kentucky, is located. He was intrigued by the story and asked Smith to see if he could acquire some teeth for him. Gist never quite made to the falls, though he came close. While approaching the falls on the Kentucky side of the river, he met two of Smith's employees, who were bringing him a pair of giant teeth. Smith's men told him that it wasn't safe to go any farther; there were hostile French and Shawnee in the area. Gist took their advice and returned to Virginia, passing the salt lick on the way without stopping to see the bones firsthand. Once back, he gave one of the teeth to his employer and kept the other for himself.

In the summer of 1755, a group of Shawnee warriors attacked the small settlement of Draper's Meadow in western Virginia. Several settlers were killed, including an infant, and five were taken prisoner. These were Mary Draper Ingles, her two young sons, her sister-in-law, and an unrelated man. The captives were split up, and Ingles was taken west to an encampment of French and Shawnee just above mouth of the same salt lick stream where Smith had seen his bones four years earlier and Longueuil had seen them twelve years prior to that. This may have been the same group that Gist was told to avoid. Ingles and another captive, known only as "the old Dutch woman," were taken to the salt lick. The two were put to work boiling water to collect salt. In October, they found two French soldiers sitting on a giant bone cracking walnuts. Ingles convinced them to give her a tomahawk, and she and the Dutch woman made their escape. It took the two women forty days to walk back to English civilization as winter was falling. Ingles told the story of her captivity and escape many times during her long life. After her death, one of her sons published her story, which included the reference to the giant bone the French soldiers were casually using as a seat. Since then, it has been retold by historians, novelized, and made into movies.

By the early sixties, the mysterious bones were becoming fairly well known in the British colonies. In 1762, another group of Shawnee, who had been allied with the French during the war, brought a tusk and a tooth to the commander of Fort Pitt as a peace offering. That same year, James Kenny, the manager of a trading post owned by the Pennsylvania government, wrote "that this Continent Produces Elephants, as Large Teeth have been found in a Lick down ye Ohio between 4 & 6 pound weight, one of which I seen Weigh'd, which Weighed 4 ¼ lb." The following year, he reported that he recorded the alternative theory of Benjamin Sutton, a trapper who have been to the lick and measured some of the bones. In Sutton's opinion, the bones came from "the Rhinosses or Elephant Master, being a very large Creature of a Dark Colour having a long Strong horn growing upon his Nose (with which he kills Elephants)." There were others. Important men in Philadelphia, New York, and London heard these stories and received teeth as gifts. They wanted to know more. The man best placed to answer their questions was George Croghan, an Irish trader, land speculator, and, arguably, the most influential white man in the Ohio valley.

When Gist visited Smith during his surveying trip in 1751, Croghan was his guide. At the time of that trip, Croghan had been operating in the valley for almost ten years. He built the trading post that Smith operated out of, he organized many of the Indian nations in the region against the French during the war, and he developed important relationships with merchant houses back east. By the end of the 1750s, his influence in the valley was such that he was given official positions representing both the Pennsylvania and London governments with the Indians. Croghan must have known about the salt lick soon after arriving in the valley, but the earliest surviving record connecting him with giant bones is from 1762. Peter Collinson, a wealthy wool merchant and member of the Royal Society, was one of those Englishmen who had had his interest piqued by the Ohio teeth. In June, he wrote to John Bartram, one of the most famous scientists in the colonies and co-founder of the American Philosophical Society, asking him to contact Croghan for more information. Collinson explained to Bartram about the object of his curiosity: "Their Bones or skeletons are now standing in a Licking place not far from the Ohio of which I have Two of their Teeth. One Greenwood an Indian Trader & my Friend Geo:Croghan both saw them & gave Mee relation of them." Collinson hoped to get a better description of the rest of the animal, especially the feet and "horn," to help "determine their Genus or Species." The two teeth he owned had been collected by Greenwood and sent to him by the governor of Virginia. In another letter, written a few weeks later, Collinson told Bartram that he also written to "B:Franklin," another friend of Croghan's, about his request. Between Bartram and Franklin, Collinson hoped for a quick answer to his questions. Collinson would have to wait. Croghan was busy unsuccessfully trying to keep the peace between the British and the Native American nations unhappy with the shift from French to British power in the valley. It would be three years before Croghan could respond to Collinson's request.

When Longueuil brought his bones to Paris and donated them to the Cabinet du Roi, Buffon had only been made its director for a year. Throughout his career, Buffon worked to transform the king's botanical garden and cabinet of curiosities into a major museum and research center. To this end, he hired Louis-Jean-Marie Daubenton as an anatomist to assist him in writing his massive encyclopedia of nature, *Histoire naturelle, générale et particulière,* and help cataloging the various collections. In both of those roles, Daubenton had cause to examine the Ohio bones. He was probably present when Fabry visited the museum and told Buffon about bones and local legends. The first indirect mention of the bones occurs in 1761 in the chapter of the *Natural History* on lions, "The prodigious *Mahmout* . . . no longer exists anywhere, although its remains have been found in several places, at great distances from each other, like Ireland,

Siberia, and Louisiana." The next year, Buffon began working on the volume that included elephants. In preparation for this, Daubenton gave the Ohio bones their first really rigorous examination.

Daubenton presented the results his research to the academy in August 1762 in a paper titled "Memoire on some bones and teeth remarkable for their size." In this paper, he compares the Ohio bones to the mammoth bones brought back from Russia by de l'Isle and those of an Asian elephant owned by the king that had died several years earlier. As an introduction, Daubenton praises the modern science of comparative anatomy. He and his contemporaries no longer need to look to the imagination to understand mysteries like these large bones. Though his predecessors thought them to be from giants and the Ostyaks imagine them to be from an enormous mole-like creature, he is confident that he possesses the knowledge and reasoning to make the correct identification.

Daubenton's methods in examining the Ohio bones are precise and careful. His first examination is of the femurs and includes a detailed illustration by the same engraver that prepared Guettard's tooth illustration ten years earlier. He shows that the bones of all three animals have the same basic shape. The angles and curves of the different parts of the bones are the same. The attachment points for ligaments and muscles are in the same place and of the same shape. The only meaningful difference between the three is in their thickness. The mammoth and the Ohio bone are much thicker that the menagerie bone. Daubenton does not think this an important enough reason to say the bones come from different species. It is much more reasonable to conclude that the differences are due to age and sex. With only one elephant to work from, he had no other way to test his thesis. He examined the tusks closely and determined that they were all made of the same type of ivory. This leads him to the same conclusion. The tusks all come from elephants; the differences in size and curvature can easily be explained by age and sex.

Then he gets to the teeth, and here he encounters a real problem. The teeth from the Ohio in no way resemble those of the elephant or the mammoth. Daubenton couldn't write to Longeueil to ask for additional information about the discovery. He had been killed in the wars with the British several years before. The only account he had was the short description written by Fabry. To address the mystery, Daubenton looked through the collection for teeth resembling the three Ohio teeth. The closest resemblance he found was with hippos' teeth. Though the Ohio teeth were much larger, they had the same basic configuration.

One possible explanation Daubenton considered was that there might be an unknown hippo-like animal with tusks like an elephant's. Hippos do have tusks, long lower canine teeth, like those of a boar. Those, he pointed out, are normal teeth and not ivory, however. In the collections, Daubenton had a fetus of a hippo. This was well enough developed that he was able to compare the femur and determine that it was shaped nothing like the other three femurs. If he considered the possibility of an elephant-like creature with hippo's teeth, he never mentions it. His conclusion is that the Ohio bones belong to two different animals: an elephant and a giant hippo whose disarticulated bones had become mixed together. The illiterate Indians who collected the bones mistakenly attributed them to the same animal. He also made a passing mention of the Theutobochus teeth as an example of how things can be misinterpreted.

Daubenton's paper was published in the 1764 issue of the academy's journal. In that same year, the volumes of Buffon's *Natural History* that dealt with elephants and hippos were published. For every animal in the *Natural History*, Daubenton wrote an anatomical essay and described the items in the royal collections related to that animal. True to his previous conclusions, he cataloged the femur and tusk from the Ohio with the elephant and the teeth with the hippo. Along with the Ohio teeth and real hippo's teeth, he listed four other fossils collected by thermometer pioneer and all-around smart guy René Antoine Ferchault de Réaumur in the early years of the century. Most of Réaumur's teeth came from *Gomphotherium angustidens*, a relative of the mastodon that lived in the south of France and Spain several million years ago. Réaumur collected the teeth, but not out of a paleontological interest. He collected them because they appeared to be made of turquoise. At the time, the composition and origin of turquoise was very badly understood. In Languedoc, a type of turquoise was manufactured by baking fossil teeth that they mined in a strip of land near the village of Simore. Once heated, the partly petrified enamel turned the exact same color and the same hardness as mineral turquoise. Since most mineral turquoise came from Persia and no Westerner had ever seen the mines, there was no reason not to believe that it was not manufactured in the same way. The Réaumur teeth and the Ohio teeth, both types of mastodon, outnumbered the real hippo teeth in the royal collections.

That same year, Croghan was in London pursuing some business interests. It's not known whether he visited with Collinson while he was there. What is known is that he remembered his request for more bones and teeth and tried to fill it as soon as he was back in the Ohio valley. For two years, the Ohio valley had been the site of a war with British settlers and soldiers on one side and a confederation of Indian nations north of the Ohio led by the Ottawa chief Pontiac. The war had been one of massacre and counter-massacre, with thousands on both sides uprooted from their homes. Because of his long residence there, Croghan was trusted more than most British agents. The crown called on him to soothe the feelings of Indians along the Illinois River after the war. It was not an easy task. Before he could get to the Ohio, his expedition was attacked by white vigilantes who opposed his peace mission. They burned most of his cargo of trade goods and gifts. Nevertheless, he gathered what was left of his cargo and journeyed on. True to his promises, he stopped at the lick and gathered bones for Collinson. A week later, he was attacked by a group of Kickapoo Indians. His boats, with the bones, were sunk. Five of his men were killed and the rest of them taken prisoner. Croghan, though having taken a tomahawk to the head, was able to negotiate with the Kickapoo leaders and gain freedom for himself and the rest of his crew. Some French Canadian traders resupplied the survivors so that they were able to make their way back to Pennsylvania. The trip wasn't a total failure. Before the disaster, Croghan was able to negotiate peace with the Indians along the Wabash River. This was enough success to make him a celebrity in the east and to maintain his credibility with the government in London.

The next year he tried again with a much larger party. The frontier was quiet enough that some friends and businessmen came along. Once again, Croghan had the expedition stop at the salt lick to collect bones. Harry Gordon, one of the military escorts, came along "to view this much talked of Place." Gordon was not disappointed. He wrote: "on our Arrival at the Lick which was about 5 Miles distance South of the River, we discovered laying about many large Bones, some of which are the exact Pattern of Elephants Tusks, & others of different parts of a large Animal." The travelers spent the whole day exploring the lick and selecting bones. Croghan and George Morgan each made large collections. This time, the expedition went smoothly. After negotiating and making gifts along the Illinois, they continued down the Mississippi to New Orleans and sailed to New York arriving at the beginning of 1767. Once there, Croghan made a public showing of the bones before dividing them up and sending one batch to Lord Shelburne, the minister in charge of the colonies, and the other to Ben Franklin, his partner in his latest land speculation scheme. Shelburne donated his bones to the British Museum, where Collinson and others eagerly examined them.

The first discussions of the bones among learned men were in letters. As soon as the bones arrived, Collinson wrote to Bartram in Philadelphia and Buffon in Paris, expressing his excitement. Ben Franklin was in London at the time giving Collinson and others the opportunity to examine the entire collection that Croghan brought back from the Ohio and compare them to Sloane's collection of mammoth bones, which were also in he British Museum. In letters to Croghan, Franklin mused about the identity of the animal. As with Daubenton, he immediately saw that the primary problem was how to reconcile the teeth with the tusks. He wrote, "The tusks agree with those of the African and Asiatic in being nearly of the same form and texture . . . But the grinders differ, being full of knobs, like the grinders of a carnivorous animal; when those of the elephant, who eats only vegetables, are almost smooth." Soon after the fossils arrived, Franklin made a short trip to Paris, where he was able discuss them with French savants. One of his correspondents was Abbé Jean-Baptiste Chappe d'Auteroche, who had recently been in Siberia to observe a transit of Venus and who had brought back some mammoth bones of his own. Abbé Chappe was fascinated by what Franklin told him and asked him to acquire a tooth or piece of jaw from the Ohio valley animal to study. In the beginning of the New Year, Franklin sent him a tooth from his own collection.

After all the dangers braved to get the bones to him in Philadelphia, Collinson was finally able to digest the results and was the first British savant to get some comments into print. On November 26, he read a short letter to the Royal Society. All he had were questions, which he submitted to the membership for discussion. Now that he had some tusks to examine, he said that he "would not hesitate to conclude that they belong to elephants." This raised numerous difficult questions. Why have no elephants' teeth been found at the lick along with the elephants' tusks? How did the bones get there? There are no elephants in North America, nor could they survive in that climate. Many have claimed that the elephants' bones found in Siberia were washed there by the Deluge. But America seemed too far from the native lands of elephants for even this extraordinary method to be sufficient. As a postscript, he points out that teeth brought back from Peru the previous year bear some resemblance to the Ohio teeth. Two weeks later, Collinson was back before the society with two excellent etchings of a tooth and some further comments. To check his earlier conclusion about the tusks, he had visited a warehouse where fresh ivory from Africa and Asia was for sale and confirmed that the Ohio tusk is real ivory. His second observation was that the teeth were not necessarily those of a carnivore. An animal of that size would be too large to be an effective hunter. Teeth with high cusps like these could have been used to break up branches for food. For Collinson, the basic question remained, are these the bones of an elephant and some large, unknown animal intermixed, or are they those of some strange, unknown hybrid with some of the characteristics of an elephant and some unique to itself. By the time Franklin got around to sending a tooth to Abbé Chappe, he had become convinced that there was nothing contradictory in the knobby teeth belonging to a specific type of elephant.

The next to publish was Dr. William Hunter, a Scottish anatomist and member of the Royal Society. Being a medical doctor, Hunter was better qualified than Collinson or Franklin to opine on matters of anatomy. Hunter had been interested in the stories of mysterious large bones from different parts of the world but hadn't made any serious study of the problem. This changed when he heard about the arrival of Croghan's bones. He arranged for the loan of a tusk and a tooth from Shelborne's collection. Like everyone else before him, Hunter looked at the tusk and concluded that it was, indeed, from an elephant. Not quite sure what to make of the tooth, Hunter showed it to his brother John, who was also a doctor. Without hesitation, John informed him that the tooth could not have come from an elephant. The nobs and the coating of enamel showed that the animal was either carnivorous or omnivorous. This made Hunter doubt his conclusion about the tusk. He next made a trip to the British Museum to examine their collections. Here he confirmed that the tusk was the same as that of Asian, African, and Siberian elephants, but that the tooth bore no resemblance to any kind of elephant. Hunter next visited Franklin to examine his collection. He left Franklin's home convinced that the tusk and teeth came from the same animal and that it was an enormous carnivore. Hunter interested Shelburne in doing more research and prepared a questionnaire to be sent to the colonies. The questionnaire mostly dealt with getting exact descriptions of the places of discovery. Like his Russian counterparts, he wanted to obtain a complete skeleton. If that was not possible, the bones he most wanted were a skull and a foot. Finally, he made a second trip to the museum to compare the Ohio teeth with other animals, including hippos and elephant-like remains from South America. Hunter challenged Daubenton's conclusion that the elephant, mammoth, and Ohio valley animal were all one species. Based on Daubenton's and his own observations, he was inclined to say the mammoth and Ohio valley animal were one species and the elephant another. It was a revolutionary claim.

By the middle part of the eighteenth century, the learned men of France, Britain, Germany, and Russia had reached a consensus that the Siberian mammoth, elephant-like bones found in northern Europe, and modern elephants in Asia and Africa were all one species and they were quite comfortable with this consensus. Prior to the discoveries in the New World, the Republic of Letters felt that the remaining problem of the mammoth was explaining how the bones got into places where elephants could not possibly live. But the discovery of the mastodon in the Americas turned this consensus on its head. The burgeoning societies of scientists in Europe had to look at the bones in Siberia and Europe with new eyes.

CHAPTER 7

SIBERIA AND PARIS

One of the great projects of the Enlightenment was the *Encyclopédie*, written by over a dozen authors under the editorship of Denis Diederot. It began as a simple translation of the very practical two-volume Chamber's *Cyclopaedia, or an Universal Dictionary of Arts and Sciences*, but it grew to include essays by some of the best-known minds of France, including Daubenton, Voltaire, Montesquieu, and Rousseau. The first edition had seventeen volumes of text and eleven volumes of illustrations. Later editions would eventually grow to 166 volumes. It was the first endeavor of its kind, and to have a repository of information in one "place" was a revolutionary undertaking and achievement for intellectual culture, although some did not see it that way. The initial expansions included science and just enough philosophy to be considered dangerous by the conservative establishment. Although the Parlement of Paris condemned the book, the editors had powerful protectors at the court. They worked around a ruling prohibiting the publication in France by writing "Neufchastel, Switzerland" on the title page and continuing on as if nothing had happened. The first volume was published in 1751 and the final volume of illustrations in 1772. Despite it being extensive and expensive, the *Encyclopédie* was one of the bestselling books of the century. Three entries directly or indirectly refer to Siberian mammoths: Behemoth, Mammoth, and Fossil Ivory. A fourth entry, "Fossil Bones," describes large bones found in Western Europe, including the Tonna and Cannstatt discoveries and the probability that they came from elephants. For the possibility that they might be the related to Siberian ivory, the reader is referred back to the "Fossil Ivory" entry.

There are two entries with the heading "Behemoth." The longer of the two is about the Behemoth of Job. It examines the various ideas about the animal that the word it could refer to and dismisses the musings of "Talmudic Doctors" as fairy tales. The second "Behemoth" entry is labeled "Natural History" and consists of two long sentences about the animal of the far north which is claimed to be the source of a beautiful type ivory. The Turks and Persians use it to make sword handles. The description is very similar to that of Father Avril sixty-five years earlier. The "Mammoth" entry is also a small one and equally out of date. It describes "Mammoth" as the name for large bones found along rivers in Siberia, which some people believe come from elephants. It misstates Gmelin as saying they look like ox bones. The "Mammoth" entry refers the reader to the "Fossil Ivory" entry in the previous volume for more information.

The "Fossil Ivory" entry is considerably longer than all the others combined but contains the same errors as the "Mammoth" entry. This not surprising as it was added a few months earlier and is probably the source referred to when creating the later "Mammoth" entry. The author, Baron d'Holbach, writes of Gmelin in flattering terms, calling his narrative of his travels excellent, but once again accuses him of saying the tusks come from oxen. He then corrects Gmelin by saying there are bones of two animals found in Siberia and that the ivory comes from the one that is not an ox, which is exactly what Gmelin wrote. D'Holbach's approach to the word is a little confusing. In his entry for "Mammoth," he states the word refers to the ivory. Here, because the entry is titled "Fossil Ivory," he generally uses the word in that sense, except for a paragraph on native beliefs. The Yakuts, he says, believe the ivory comes from an animal they call *mammon* or *mammut.* He follows this with the usual short mention of the "fabulous" underground creature that they believe in. There is a breath of irony in his confusion over the word. Nicolaas Witsen used the word in both senses and very much like d'Holbach does, first saying it is the name of the ivory, then saying the Siberians believe it comes from a beast of that name. Johann Bernhard Müller, in his *Life and Customs of the Ostyaks*, was the last to use the word as the name of the ivory. D'Holbach's use of it that way was based on scholarship a half century out of date. However, two centuries later he would be vindicated in his use. In 1976, Michael Heaney published a study of the likely source of the word in Siberian languages and determined that it came from two words in the Mansi language meaning "earth horn."

D'Holbach admits that the ivory resembled elephant ivory and touches on the perplexing problem of getting these elephants to Siberia. He looks at the possibility, mentioned by others, that the armies of some would-be conquerors brought them and rejects it. While the most common objection to this idea is that there is no record of such a thing, d'Holbach is blunter and asks why any ancient would want to go to the ends of the earth to battle Scythians and Hyperboreans. He then goes in the more daring direction of claiming these were a different type of elephant, native to Siberia during an unrecorded era when that part of the earth was warmer. Some revolution of the globe must have changed the climate in the ensuing centuries and buried the thousands of dead elephants. There are two things that he does not say in describing that scenario. He does not use the dreaded word "extinct." He hints at it, but openly stating it is still more than he will risk. Nor does he claim the cataclysm was the biblical Deluge. Earth history had become secularized enough by the 1760s that he didn't feel obliged to mention it.

Daubenton, who authored many of the entries on natural history, might have been a better choice to write more current "Mammoth" and "Fossil Ivory" entries, but it wasn't to be. A third contributor had an interest in mammoths. This was the great Voltaire, who was a regular correspondent of the most influential person in Russia: the empress Catherine II, or Catherine the Great. In the first edition of his history of Russia, he repeats a part of Avril's description calling the source of the ivory a Siberian animal, larger than the Nile crocodile, that is hunted for its teeth. He rejects that the ivory could have come from elephants, mentioning only foreign armies as a means to get the ivory into the Arctic. In the second edition of his history, he writes that he has been corrected and that that the animal is the walrus. He doesn't mention, or isn't aware, that there were two types of ivory supplying the Siberian trade and that walrus ivory was only a small part of that trade.

When Gerhard Friedrich Müller and Gmelin wrote their memoirs of their time in Siberia, both mentioned the trade in mammoth ivory. Müller's account was a formal description of the importance of the trade to the Russia Empire. Gmelin's was more of a passing anecdote. In Yakutsk, he saw Cossacks going out for the express purpose of hunting for ivory, though he suspected many were using that as an excuse to engage in other trade. This was a change from the time of Messerschmidt and the Swedish POWs. Lorenz Lange and others mentioned that some of the lower-ranking Swedish officers in Tobolsk made a living carving mammoth ivory for the China trade. However, no one mentioned the existence of ivory hunters making a systematic effort to supply that industry. It's most likely that collecting ivory was something trappers and traders did when the opportunity presented itself, but not something they made a special effort to do. Just twenty years later, things had changed considerably. When Steller attempted to go to the Arctic coast at the mouth of the Kolyma River to hunt for mammoth remains, he was not following a random hunch; he had definite intelligence that that was the best place to hunt for ivory. His intelligence was good. The region between the Kolyma and Lena Rivers was so rich in fossil ivory that some nineteenth-century maps labeled it "the Mammoth Coast." After 250 years of exploitation, it's still one of the best places to "hunt mammoths." What had happened over the intervening years was thanks in large part to Peter the Great's immense curiosity. In 1722, he issued an order that should anyone find ivory, they should search the area for the rest of the skeleton and send the bones to the Kunstkamera. His successors left this law on the books. With this, following up on reports of mammoths went from an act of idle curiosity to an imperial duty made attractive by the existence of a bounty and lucrative commercial possibilities. By the middle of the century, Yakutsk had a formal ivory market and was the main export center in the empire. In 1840 Alexander Middendorff estimated that about 110,000 pounds of ivory were exported every year and that the tusks of 20,000 mammoths had been sold since the conquest of Siberia. This systematic collection of ivory and bones greatly added to the materials available for scholars to examine. Despite a fire in 1758 that destroyed many items in the natural history collection, the Kunstkamera had three mammoth skulls and "countless" tusks to work with in the late 1760s.

By the mid-1760s, most of the first generation of members of the Russian Imperial Academy were gone. Some died in Russia. Some stayed for a few years or decades, then returned to the homes of their youth. Only Müller remained year after year, watching members come and go. In 1765, after forty years at the academy, he abruptly moved to Moscow to open a new model school. Since the death of Peter the Great, various attempts had been made to create a proper system of primary education to provide Russian-born civil servants for the empire. The results had been mixed. Now it was Catherine the Great's turn. Like Peter before her, Catherine was eager to modernize her realm and cultivated relationships with major thinkers in the West.

The plan for the school came from Ivan Betskoi, her adviser on educational matters. Betskoi envisioned a system that would educate both well-born girls and promising boys. The boys' school would be open to both the well-born and foundlings (orphans and abandoned, illegitimate children). He originally offered the job of establishing the boys' school to Anton Büsching, the director of the Lutheran school in Moscow. Büsching politely declined—he was making plans to return to Germany—but recommended Müller. Müller was an excellent choice. He was one of the few academicians who took his educational duties seriously; he frequently wrote about educational reform. Betskoi recognized the appropriateness of the match and offered the job to Müller, who accepted. Moving to the old capital did not mean breaking his ties with the academy. He remained an academician and became the deputy for the academy in Moscow, performing functions such as preparing expeditions into Siberia. Although more than twenty years had passed since his work during the Great Northern Expedition, Müller's firsthand knowledge was still invaluable. One who greatly appreciated his experience was a man named Peter Simon Pallas.

Pallas arrived in St. Petersburg in 1767, on the direct invitation of the empress, to join the academy as a naturalist. He spent his first year in St. Petersburg, studying the collections and documents in the Kunstkamera and reading books on the natural history of his new home. Pallas had a great respect for his predecessors. During his time in the Kunstkamera, he discovered the papers of Messerschmidt. Although many of his samples were lost in the 1758 fire, all of Messerschmidt's papers were still safely preserved. Pallas thought they were of great value, and he would eventually gain permission to publish some of them along with unpublished reports by Steller. Later, he would arrange to publish some of Müller's journals in Germany.

While in St. Petersburg, Pallas wrote his first major paper: an illustrated article about Siberian skulls that he had seen in the Kunstkamera. Speaking of the mammoth skulls and tusks, he confirms what is by then the scientific consensus: the mammoth is an elephant. He reviews the wide spread of mammoth bones from France to Siberia to French and English America. He also mentions some of the theories of how the bones had been transported to the places where they are found. While it is reasonable that the Romans could have brought elephants to France and Spain, no human agency, whether Hebrew tribes or Mongol armies, could account for the sheer numbers of mammoths in Siberia, nor could it explain why most are found in the farthest north part of Siberia. Additionally, no humans would have left thousands of valuable tusks behind. He gives a nod to the fact that many people believe the Deluge or some similar catastrophe brought them there, but does not commit to supporting any theory, though the most likely cause for them being found in the far north is something like the Deluge. He must have figured that most of his readers knew what a mammoth skull looked like and did not feel it was necessary to include an illustration of one. What he does illustrate is the skull of the "ox" that Gmelin had sent from Yakutsk. Pallas corrects Gmelin's identification and correctly labels it a buffalo. He backs this up with a discussion of Indian and South African buffalo, even mentioning North American bison (a misconception that continues to this day; bison are not buffalo). The third type of skull he looks at is that of a rhinoceros. This was not the same skull Spiridon Portniagin had seen fifty years earlier. That one was never recovered. It could have come from a second unknown skull that Gmelin heard about but didn't see that was found on the Lower Tunguska River. Whatever the provenance, the illustration leaves no doubt that it is a rhinoceros. Along with the skull, he includes a plate of two long horns that he identifies as having come from rhinoceroses. Pallas does not think either the buffalo or the rhinoceros represent new species. The known buffalo and rhinoceroses show enough variation that the Siberian skulls could fit within believable parameters.

Barely a year after his arrival, he was assigned to lead his own expedition into Siberia. His was one of several who traveled to together before heading off to pursue their individual objectives. Gmelin's nephew, Samuel, journeyed into Central Asia, where he died of dysentery in the captivity of a Turkmen khan before he could be ransomed. The Abbé Chappe stayed in western Siberia to observe a transit of Venus across the sun and collected mammoth bones while he was there, leading to his later conversations with Ben Franklin. Müller gave Pallas the instructions he had written for Krasheninnikov in 1737. These were the same kind of instructions most of the academics had carried during the Northern Expedition: go deep into Siberia, look at everything.

The academy accepted his paper on fossil skulls and published it in their journal a year after Pallas left for the East. As fine as the paper was, it would have been much more impressive if the academy had waited before publishing it. After leaving the younger Gmelin and Chappe, Pallas, a painter, and three other naturalists forged their way deeper into Siberia. He writes of eagerly anticipating the "greater miracles" the land had to offer. Mammoths were on his mind, and he reported that "from the Don to the northeast corner of Asia, there is not a river, especially among those that flow through the plains, that does not have bones of elephants or other animals foreign to this clime along its banks." He arrived in Irkutsk an hour before midnight on March 14, 1772. The horses, he writes, were tired. He knew the city held curiosities that he wanted to see and stories that he wanted to hear about the unknown lands across Lake Baikal. Irkutsk did not disappoint. Governor Adam de Brill told him that he had the preserved parts of an unknown large animal that Pallas immediately recognized as a rhinoceros. He was exceptionally lucky that almost everyone involved in bringing the animal to his attention had understood its importance. The rhino had been discovered by a group of Yakut (Sakha) hunters in December on the banks of the Vilui River, the last major tributary of the Lena, which falls into it well above the Arctic Circle. The rhino was nearly complete when they found it, but enough of it was in a bad state of decay that they decided to cut the feet and head from the carcass and leave the rest behind. Even if they had wanted to retrieve the entire body, it would have been almost impossible to dig the unexposed parts out of the frozen ground in midwinter. The hunters took the good parts to Ivan Argunov, the district magistrate, who took a notarized statement from them detailing the location and position of the carcass and then sent the parts and statement to Yakutsk. The authorities there kept one foot and sent the rest on to Irkutsk, where de Brill received them just three weeks before Pallas's arrival.

The head and feet were in excellent condition. The delicate structure of the eyelids remained. Muscles and fat were preserved under the skin. Though the horns were missing, from the spots where they had been attached, he could tell it had been a two-horned rhino. Almost all of the skin was present and thickly covered with hair. All varieties of rhinoceros known in his day had very sparse hair, so he knew this one was something special. Of immediate concern was making sure it remained preserved in the best condition possible. It had already begun to give off a stench that he compared to "an ancient latrine." He chose to dry it in an oven. The melting fat falling in the fire caused the oven to get too hot, and one of the feet was burned beyond any hope of saving. Naturally, the servants watching it were blamed, even though they had no experience roasting a rhinoceros. Before sending the parts to the academy, Pallas took careful measurements and wrote a detailed article for the academy. He regretted not having had time to have his artist prepare drawings of it, but the academy made up for this by having their own artists prepare excellent illustrations for the journal. Almost eighty years later, Johann Friedrich Brandt would publish an article on Siberian rhinoceroses in the academy's journal that would include images of the head that, despite decades of improvements in printing technology and color, were not as good as the earlier ones. Both

Ides's traveling companion and Messerschmidt's informant, Wolochewicz, witnessed mammoth remains with soft tissue attached, and Siberians spoke generally of bloody mammoths being found, but this was the first time the flesh of an ancient frozen animal had been recovered. It would be commented on and carefully studied well into the next century.

Among the first outsider to write about the discovery was the Dutch anatomist Peter Camper. Camper is primarily remembered for his efforts at racial classification, which were a precursor of the scientific racism of the nineteenth century. This ugly legacy has overshadowed his other accomplishments, including those in natural history. The year before Pallas's description of the Vilui rhinoceros was published, Camper received the head of a rhino killed in southern Africa and dissected it. Camper wrote a monograph comparing various rhino specimens, both African and Asian, that was published in 1782. Five years before that, he sent to Pallas a short description and drawings of the African head to be used for comparative purposes. Pallas had the paper published in the academy's journal along with comments of his own. Throughout his description, he points out details that differ from the Siberian head and suggests points for further investigation. In his conclusion, Camper says that he doesn't think they were the same species and that he desires to dissect an Asian rhino for further comparison. He finishes with a meditation on fossil mammals, mentioning the "Ohio animal(s)" and elephants found in the gravels of his home country. Pallas's comments mostly focused on teeth. Because the published drawings of the Vilui rhino showed it with its skin attached, Camper hadn't been able to make those comparisons. In his note, Pallas mentions that he has a more recent letter from Camper describing his examination of a live young rhino at Versailles. Their conclusion is that rhino teeth are very confusing and the subject needs more study. Camper was already collecting bones for that purpose and would get the complete skull a Javan rhinoceros to add to his comparisons ten years later. Camper was also interested in elephants and hoped to get bones from mammoths and the Ohio animal to examine. The latter proved harder to collect than rhino skulls.

When Croghan sailed down the Ohio and Mississippi Rivers with the mastodon bones that would stir so much interest in Britain and France, he had with him George Morgan, who collected a second box of bones. Back in New York, Morgan gave them to his brother, Dr. John Morgan. Dr. Morgan had a small collection of natural history curiosities that included a hairball from a cow and some bits of lava from Mt. Vesuvius. Despite the rarity of the bones and the public interest in them, Dr. Morgan treated them with surprising indifference. Years later, in 1783, Christian Friedrich Michaelis was shocked to find them unceremoniously piled in a corner still covered with dirt from Big Bone Lick. At the time he visited, Michaelis was hunting for some good mastodon bones to send to his father in the Netherlands. Michaelis offered to clean the bones, and Morgan gave him permission to borrow them for the purpose of have detailed drawings made. The artist who did the work was Charles Willson Peale. While Peale worked, several influential people made special trips to his shop to view the bones. This gave Peale the idea to open a natural history museum of his own. Michaelis, meanwhile, had been bitten by the fossil bug and sought out people who had been the Big Bone Lick and other fossiliferous places on the Ohio to learn as much as he could. He was a credible observer of anatomy. He had come to America during the revolution with Hessian mercenaries as an army doctor and stayed after the hostilities ended. He had excellent knowledge of human anatomy, but he was not that good with animal anatomy.

Michaelis was already acquainted with Camper through Michaelis's father. Because Michaelis had handled so many bones of the Ohio animal, Camper was inclined to defer to his authority. Some of Michaelis's conclusions were solid. He rejected Daubenton's idea that the American bones were those of elephants mixed together. He also rejected Hunter's idea that the teeth indicated a carnivore. He believed that the bones and teeth represented one unknown animal. These were not particularly controversial opinions; it was one final assertion that got him in trouble. One of the bones in Morgan's collection was a piece of an upper jaw consisting of the palate and three teeth. Michaelis looked at it backward, thinking the back was the front. From that angle, he decided that there was no room for tusks on the animal. For no particular reason—he hadn't seen the front of a skull—he also decided it hadn't had a trunk. And if that wasn't enough, he declared that it and the Siberian mammoth were the same tuskless, trunkless animal, meaning there was no explanation for the many tons of ivory coming out of Siberia each year.

On his return to Germany, Michaelis sent Camper a bone and a set of Peale's drawings. Camper had already spent some time studying the Ohio animal and owned a tusk of one. He had visited the British museum and studied their bones. Through his study, he had come to a series of correct conclusions about the animal: it was a separate species from the elephant and the hippo, it had tusks and a trunk, and, like Franklin, he saw that the teeth were suited to a browsing herbivore. Despite his superior knowledge, Michaelis completely convinced him, and Camper even went so far as to publish a paper recanting his former views. Despite this, Camper wasn't done with the topic. In 1784, he offered to buy Morgan's collection so he could examine them himself. Morgan refused, having decided they should stay in the country where they had been found. Three years later, a visitor from Philadelphia, Samuel Vaughan, convinced Camper to make another offer. This time, Morgan agreed. He was old, he wrote, and would never make a detailed study of the bones himself. He was impressed by Camper's reputation and knew he would do them justice. Vaughan oversaw the details of shipping the bones. By now, Camper was also an old man. He died just weeks after the bones arrived in Amsterdam.

When Pallas traveled in Siberia between 1768 and 1774, he brought something to that part of the world that none of his predecessors had; he was a geologist in the modern sense of the word. Messerschmidt, Tatishchev, Strahlenberg, Müller, and Gmelin had all studied minerals, but they did so with an eye toward locating exploitable metal ores, instead of studying minerals within the context of the earth itself. The word "geology" was still new, and Pallas was the first to make an extensive study of the landforms of Siberia. In particular, he studied the Ural and Altai Mountains and developed a general theory of mountain chains. He published the first draft of his theory at the same time he was corresponding with Camper about rhinos and a slightly updated version in 1782. The cores of mountain chains, he writes, are great masses of granite that were once islands in a world-covering ocean. The peaks, where the granite cores are most exposed, existed before life appeared on earth. As we move down slope, the land grows younger and maritime fossils begin to appear. Finally, on the plains, fossils of plants and land animals become common. The modern world was not formed only by erosion of these first mountain cores. Periodic massive volcanic events raise major islands and chains. These are so sudden that they displace entire seas, which flood across the adjacent dry land. One such event could have created the Philippines and Japan and swept elephants, rhinoceroses, and buffalo into cold Siberia, where they were buried by sediment and frozen. Pallas's theory was a grander version of Steno's and Leibniz's theories, and it gave the earth a complicated history that took time to unfold. He avoided making a definite statement about the age of the earth, but the implication that it was older than five to seven thousand years was inescapable.

When Buffon began to write his *Natural History* in 1749, he prefaced it with a volume describing the history of the earth up to the advent of humanity. His theory covered a broad scope in which uncounted thousands of years of geological change occurred, with humans only showing up at the end. The French religious authorities were not amused. Four years later, he was compelled to issue an apology and recant his position in the introduction to the fourth volume. Twenty-five years later, in 1778, his own authority as the most respected thinker in Europe had enough weight that he was able to issue a new and improved version of his theory without official opposition. This new volume, called *Epochs of Nature*, divided earth history into seven epochs. Humanity inhabited only the last period. In the first epoch, a comet grazed the sun, throwing a small part of its mass into space. During the second epoch, globs of this mass cooled to become the planets. During the third, the glob that was the earth cooled to the point where water precipitated and covered the whole surface. The first life, in the form of shellfish, appeared and covered the ocean floor. During the fourth, dry land appeared as part of the oceans drained into deep caverns in the earth, revealing seashell-covered land, and volcanoes threw up great mountain ranges. During the fifth epoch, plants and land animals appeared in the far north, including Siberia. Prominent among these lifeforms were elephants and rhinoceroses. These were so important to his theory that the chapter about the fifth epoch is titled "When elephants and other animals of the South inhabited lands of the North."

Buffon used the observations of Steno, Leibniz, Pallas, and others to put together a generalized global sequence of strata, which addressed the presence of maritime seashells on mountaintops and far from the seas. These were laid down in primal seas of the third epoch and later thrust up with the creation of moutains during the fourth. In his scheme, heat was the key to the formation of the earth and to the distribution of life upon it. Buffon believed life was an emergent quality of matter and would appear wherever and whenever conditions allowed. Life appeared first in the seas once the earth cooled to the point that water precipitated on the surface. Plants and land animals appeared in the farthest north, because that was the area of the earth that cooled first. The first animals were tropical in nature. As the earth cooled and the habitable zone moved south, these tropical animals moved south with it. New lifeforms appeared to occupy the cooler, vacant regions north of the receding tropics. These lifeforms were inferior and possessed less vigor than their tropical predecessors. Buffon felt this narrative nicely explained the vast amounts of ivory coming out of Siberia. The problem of explaining it had become more acute since Ivan Lyakhov had explored the islands that now bear his name. Lyakhov first visited the islands in 1750 and noted that there was so much ivory that it was as if the islands were made of it. After Catherine came to power and granted him a monopoly on exploiting the islands, his agents began to bringing enormous amounts of the precious substance to market. Buffon thought it was now impossible to explain the presence of elephants in the north by any one-time occurrence like stray armies or even the Deluge. For him, the only possible solution was that elephants had inhabited Siberia for a very long time.

Buffon was prepared to take a somewhat qualified stand over the age of the earth. When he was still simply Georges Leclerc, he purchased the village of Buffon, which his father had once owned. There he built an arms factory that made him a wealthy man and that provided him with a workshop to pursue his scientific interests. To test his idea of the cooling earth, he had his workshop heat various sized iron spheres (cannonballs) until they were red hot. He then timed their cooling until they were, first, safe to touch and, second, cool enough for things to grow on them, in his opinion. He does not record who did the actual hot cannonball touching or if they received a bonus in their paycheck for the job. After he was satisfied that he knew the rate of cooling for iron, he experimented with mixed materials that he thought better approximated the actual makeup of the earth. Once he had completed his experiments, he arrived at the astonishing conclusion that the earth must be three million years old. Though he was no nonger afraid of the religious authotities, he feared this was too much to ask his scientific colleagues to accept. Buffon agonized over these numbers for years before publishing *Epochs of Nature.* When he finally did publish, he reduced all of his estimations to their lowest possible values and estimated the age of the earth at 75,000 years. Even so, this was ten times older than the largest religious estimate.

Buffon's theory also provided a means to put mammoths and hippos in North America, although he still held on to his and Daubenton's identification of mastodon teeth as belonging to giant hippos. Buffon wrote that violent actions of the fourth epoch that had drained away most of the original ocean and thrust up the great mountain chains didn't end with that epoch; they had only become less common. To account for the modern distribution of species, Buffon also theorized that the configuration of the continents had changed since they first appeared. With the possible exception of South America, they had all been connected at one time making it easier for the first animals to move south. Before Gibraltar and the Bosporus opened, the Mediterranean was a series of lakes that elephants could easily pass on their way from Europe to Africa. He also believed that while it was likely that the Aral, Caspian, and Black Seas had once been connected, draining part of the water into the Mediterranean, and evaporating another part cleared the way for another group of elephants to migrate to India. For North American elephants, Buffon's problem wasn't so much getting them there—the northern parts of the continent reached the same latitude as Siberia—as separating North America from Eurasia. The great caverns into which much of the original ocean had drained were subject to occasional collapses. He suggested several regions as candidates for new seas created this way. One was the western Pacific. Another was the Bering Strait. A third was the entire North Atlantic. The Bering route would have been the easiest way to get elephants to Ohio, but he was rather fond of an Atlantic route from Europe to Britain to the Faroe Islands,

Iceland, Greenland, and Canada. He also suggested a southern route to Florida from Spain, to Atlantis, the Azores, and the Greater Antilles. Although he doesn't mention it, the northern Atlantic route was the same as the one suggested by Molyneux to get the Irish elk, which he thought was the same as a moose, to the British Isles.

Buffon's theory of the earth was not widely embraced, even though, to our modern minds, he was dancing around the right ideas with regard to continental drift. Unfortunately for Buffon, grand theories such as his had gone out of style in favor of specific research that produced measurable results. But though it was not widely embraced, it did not receive wide censure, either, minus the initial attack from the church authorities. His theory did, however, grievously offend one person: Thomas Jefferson. Jefferson wasn't particularly upset by the geological parts of Buffon's theory. It was the fact that Buffon wrote that new lifeforms that appeared in the north were inferior and possessed less vigor than their tropical predecessors. This wasn't a result of appearing later; it was because they lived in a colder climate. This inferiority wasn't limited to later developing lifeforms per se. Superior, older breeds would degenerate if they moved to cooler climates. Between the first and second presentations of his theory of the earth, he developed this idea and specifically pointed at North America as an example of this degeneration. Buffon made his most offensive comments in the ninth volume of *Natural History* (1761) following the article on the lion. One essay in the volume, "Dissertation on Animals Peculiar to the Old World," points out the grand animals of the Old World that are not to be found in the New. America has nothing that can compare to the elephants, rhinoceroses, and tigers east of the Atlantic. In "Dissertation on Animals Peculiar to the New World," he claims the Americas have few unique animals. Most that are there are degenerate forms of Old World animals. The few that are unique are noxious animals like snakes, insects, and frogs. In Buffon's imagination, the entire New World was one environment, a cold, damp forest with wild rivers and few signs of human enterprise. While the Europeans in America grew weak and lazy, the natives showed the final end of human degeneration: "In the savage, the organs of generation are small and feeble. He has no hair, no beard, no ardour for the female. . . . Their heart is frozen, their society cold, and their empire cruel. . . . They have few children, and pay little attention to them. They are indifferent, because they are weak." In the following years, Buffon toned down his criticism and, by the time he wrote *Epochs*, used only South America to illustrate degeneracy. But the damage was done. *Epochs* was not read as widely as the earlier volumes, and, in the intervening years, the idea of American degeneracy—for both animals and humans—had been enthusiastically and vocally embraced by many European writers.

Jefferson wasn't alone in his understandable offense. Franklin, Alexander Hamilton, James Madison, and both John and Abigail Adams all expressed their displeasure, but Jefferson was especially inflamed. Through war and revolution, he would hang on to the slight until he finally found the right medium to respond in 1780. It would lead to the only book he published in his lifetime. The opportunity came in the form of a simple questionnaire from François, marquis de Barbé-Marbois. Marbois, the secretary of the French legation in the rebelling American colonies, was assigned to gather information on the individual states with an eye toward trade opportunities. In Virginia, he presented the questionnaire to Joseph Jones, Madison's uncle, who forwarded it to Jefferson, who he rightly assumed was the most knowledgeable person in the dominion on matters of resources and natural history. Jefferson's response was slightly delayed by the British invasion of Virginia, but he was able to present Barbé-Marbois with a book-length response by the end of 1781. The longest portion of the book is his answer to query 6, which asked him to describe the "productions, trees, plants, fruits, and other natural riches" of Virginia. In it, he spends several pages describing the mammoth, by which he means the Ohio animal. Based on the traditions of several eastern Indian nations, which described it a monstrous and deadly bison, and his own rejection of extinction, Jefferson was sure that the animal still roamed in unexplored regions northwest of Virginia. He described it as being six times the size of modern elephants and therefore far greater than anything the Old World had to offer. After the war, he presented a small number of friends with handwritten copies of his book. Others heard about it and asked for copies, which Jefferson had printed in Paris, where he took up his new position as the American minister plenipotentiary.

In Paris, Jefferson very anxiously looked for an opportunity to meet Buffon and debate him. He had with him the tanned pelt of a large cougar that he hoped to present to the now quite old naturalist as a way of redeeming the honor of American cats. He had instructed his friends back home to hunt for a very large moose to have stuffed and sent to him. While he waited for the moose, he sent Buffon the cougar skin with his compliments. The French savant sent him a terse note of thanks and an invitation to dinner. Jefferson wrote that the dinner was quite pleasant and that, rather than enter into an argument, Buffon presented him with a copy of his just-published *Epochs.* After dinner, Buffon conceded that the cougar skin proved American cats were just as impressive as their Old World counterparts. When the moose finally arrived, Jefferson sent it to Buffon, who finally relented. In a note from his secretary, he promised to make corrections in future editions. There would be no future editions. Buffon died six months later.

By the last decade of the eighteenth century, Karl Linné's system for classifying living things had become widely used throughout the Western world This is not to say it wasn't controversial. Buffon, in particular, had hated it. He believed that all attempts to classify living things above the level of species were arbitrary and unnatural. Linné's system was based on a strict hierarchy that categorized plants and animals according certain characteristics. One of the most fundamental divisions was their means of reproduction. According to Linné, mammals are characterized by, among other things, live birth, as opposed to eggs. Buffon preferred a version of the ideas of Leibniz that species had a certain flexibility to adapt and change and even reverse that change, without becoming a new species. Linné defined species by structure; Buffon preferred to define them by utility. The highest category of animals was domesticated animals. Within that category, horeses and dogs held pride of place. Next came other domesticated animals in France. Domesticated animals in other parts of the world came in a distant third place.

Whether Linné's system was unnatural or not, it was useful. The Siberian explorers Messerschmidt, Steller, and Gmelin each brought back hundreds of new plant species to be cataloged. The workers at the Kunstkammera needed something like the Linnean system to make sense of these vast collections, and thus it was quickly embraced. The most important controversies about Linné's system concerned how to use it and fill it in. Which features of living things were the most important criteria to use in classification, and how should they be ranked? After he dropped the mineral kingdom from his system, Linné had very little use for fossils that couldn't be matched to living species. Johann Friedrich Blumenbach worked on all fronts of the Linnaean system: He added new species, he challenged the criteria for classification, and he added fossils to the system.

Blumenbach wrote about classifying the position of fossils after he returned from the Swiss trip during which he had tracked down the bones of the Lucerne giant, which had been examined by Felix Plater in 1584. His interest in biological categorization was apparent from the beginning of his career. His medical thesis at the University of Göttingen in 1775 was about defining subspecies of humans according to Linnean principles. Later, scientific racists would use his work in ways he never intended. He divided fossils into three simple categories: known, dubious, and definitely unknown. Blumenbach gave examples for each category of samples he had seen during his trip ro Switzerland. He had hinted at this idea in the first (1780) edition of his popular text *Handbuch der Naturgeschichte* (Handbook of Natural History). By the beginning of the 1790s, he laid out his categories and gave specific examples for each in new editions. In the dubious category, he named the cave bear, Irish elk, mammoth, and woolly rhinoceros. For a definitely unknown animal, he suggested the animal of the Ohio valley. In the tenth edition of the *Handbuch* (1799) he gave each of the animals a Linnaean binomial name. The mammoth was now *Elephas primigenius*, the primeval or original elephant. Blumenbach's decision to name the species as distinct from the elephants of Africa or India was a major landmark on the path to accepting extinction, but it is also a bit peculiar in the given context. The dubious category was meant to be a holding pen for fossils awaiting categorization into known or unknown. In in putting forth tentative names, he may have been laying claim to naming rights should the fossils be proven to be new species. Recent events in France could easily have encouraged him to want to act fast.

Less than a year after Buffon's death, a series of actions aimed at relieving a financial crisis snowballed to become the French Revolution of 1789. As revolutions are inclined to do, this one grabbed the old institutions by the neck, gave them a good shaking, and reconstituted them in a new way. State-sponsored science was no exception. In Paris, the various royal collections were combined and reorganized into the Muséum national d'histoire naturelle and placed under the leadership of a new generation of thinkers. The academy didn't fare as well. The museum was created in more moderate days after the French Republic was declared. By the time the ruling convention turned its attention toward the academy, the revolution had entered its most radical phase and the Terror was just weeks away. The academy was abolished and nothing was created to replace it. Many of the savants safely returned to their homes and waited events out, but others were not so lucky, including the chemistry pioneer Antoine-Laurent de Lavoisier, who went to the guillotine.

By the summer of 1795, the Terror was over and Paris had become relatively safe again. The surviving academicians returned to the capital, and with them came a cohort of ambitious younger thinkers. When a young man named Georges Cuvier arrived in Paris, he headed directly to the museum to seek employment and was promptly hired. Cuvier (born Jean-Léopold-Nicolas-Frédéric Cuvier) was born in a French Protestant enclave in Wurttemberg. He showed an early interest in geology and biology; he had read all of Buffon's *Natural History* before he was twelve. Though his fluency in German was an additional asset, there were no jobs available for young man of his talents near his family, but there were always jobs for tutors with the rich and noble in France. He was lucky to find employment with a relatively unimportant noble family on the coast of Normandy, where he quietly waited out the Terror while examining the local geology and bringing his geological ideas into clearer focus.

At the museum, he was assigned to be the assistant to Jean-Claude Mertrud, the professor of animal anatomy. Cuvier was at first inclined to specialize in marine invertebrate anatomy, but a victory by the revolutionary army led him in the opposite direction to large land vertebrates. At the beginning of the year, France liberated/conquered the Netherlands. As part of the spoils of war, they seized the Stadtholder's collection—the equivalent of the royal collections of the British Museum—and sent it to the museum. The new specimens included a wealth of fossils and skeletal remains as well as two live elephants. As Cuvier and Étienne Geoffroy Saint-Hilaire, another rising star, worked on cataloging the new treasures, they became excited by the many and varied rhinoceros and elephant samples they had to work with. The two published a short paper on rhinoceros species that dealt only with the samples at hand and did not mention Pallas's Siberian rhinos. In January 1796, Cuvier was elected to membership in the Institut de France, the successor of the academy. Four months later, he made a public presentation of part of his research. It was a paper about elephants.

The published version of "Mémoires sur les espèces d'éléphants vivants et fossiles" (Memoire on living and fossil species of elephants) is only six pages long. The first page carries a note that this is the abstract of a longer paper that will be published by the institute at a later date. His main argument is that there are three species of elephants, not one as is usually supposed. From the shape of the teeth, he says that the elephants of Ceylon and of South Africa and the Siberian mammoth must be three distinct species. He goes on to say that the Ohio animal differs even more from the three elephant species than they do from each other. He concludes with a short research program: investigate what kind of revolutions in the could have removed mammoths and other missing species. He explicitly rejects Buffon's cooling earth. As examples of other missing species, he gives Blumenbach's list (the *Handbuch* was well enough known that he must have read it). He adds one more animal: "the skeleton which has just been found in Paraguay."

In many ways, the skeleton from Paraguay was even more amazing than the animal of the Ohio. As Cuvier and Geoffroy were sorting their way through rhinoceros and elephant bones, a packet containing a large drawing of an unknown skeleton and some brief notes about it arrived at the museum. The drawings had been sent by Philippe Rose Roume, a diplomat passing through Madrid on his way to oversee the transfer of Santo Domingo (the modern Dominican Republic) to French control. While in Madrid, Roume visited the royal natural history collection, which had just been transformed into a public museum. There he saw the reassembled skeleton of a rhino-sized animal that resembled no animal he was familiar with. He learned that the display was the work of Juan Bautista Bru de Ramón, the chief artist and taxidermist of the Spanish Academy. Bru explained that the bones had been excavated near the village of Luján in Spanish Argentina by a Dominican friar named Manuel de Torres in 1787.

Bru had almost all of the bones. The only missing bones that he wasn't able to reconstruct from the equivalent bones on the other side of the body were those of the tail. He chose not to speculate about the length of it in his mounting. He had already prepared a large profile of the skeleton. Roume had a copy made and sent it to Paris with some notes from Bru.

Cuvier dropped everything and rushed a short paper about Bru's animal into press. Not having access to the actual bones and having only the most basic information about the circumstances of the discovery, he contented himself with commenting on the illustration. First, he assures his readers that he has no reason to doubt the accuracy of any part of Bru's reconstruction. Next, he points of those features that he feels are most important in identifying the animal. Cuvier had very firm opinions on the correct way to conduct comparative anatomy. The body of his paper is a lesson in his method. The broader grouping in which he placed the animal was one he called the unguiculates, a family that included armadillos, anteaters, aardvarks, pangolins, and sloths. The defining characteristic of this group are large claws on their forelegs used for digging. He walks his readers through the process of determining the relative importance of certain features and eliminating different species and genera until only the humble sloth remains. He recognizes that the size and robustness of the animal are enough to earn it its own genus next to the much smaller surviving sloths. He gives the new genus the unimaginative name of *Megatherium*, or huge beast.

Modern readers might want to grow indignant on behalf of Bru, whose patient and thoughtful work was appropriated by someone half his age, scooping his announcement and claiming naming rights (I originally did). Bru didn't look at it that way. He was working on his own detailed monograph and was happy to have his, frankly, more prestigious French peers give him some publicity. When he finally published his paper, he included a Spanish translation of Cuvier's article. When Cuvier later wrote more extensively about the *Megatherium*, he included a French translation of Bru's descriptions.

By the time Cuvier got around to writing the longer version of his elephant paper for the institute's journal, he had collected a substantial amount of new information that needed to be added. This time he not only refers to his own observations of the museum's specimens, he also delves deep into the literature of the previous century. At various times, he makes reference to Tentzel, Messerschmidt, Gmelin, Daubenton, Pallas, and Camper. He cites the last three as his predecessors in practicing the right kind of comparative anatomy. This is very gracious in the case of Daubenton, since, in the topic at hand, he disagreed with many of his conclusions, in particular, the number of elephant species and the identity of the Ohio animal. Pallas is the respected elder whose work he cites the most often. This makes sense because the Kunstkamera contained many fossil types that weren't available to Cuvier but were extensively studied by Pallas. But, for mammoths, Messerschmidt (who he spelled Messer-Schmid) was the most valuable. Despite collecting for more than a century and looting collections across Europe, the museum still did not own a mammoth skull. Cuvier made his determinations about the mammoth based on Messerschmidt's drawings as published in the *Philosophical Transactions* sixty years earlier. To trust another's artwork was high praise coming from him. He was an excellent anatomical artist and highly critical of others. When he said he had no reason to distrust the accuracy of Bru's drawings, he was giving a high compliment. Of the Ohio animal, he repeats his previous conclusions that it is a distinct species but close enough related to elephants that it belongs in the same genus. He gives both animals Linnaean names. The mammoth he calls *Elephas mammonteus.* The Ohio animal he calls *Elephas americanus.* When listing other unknown species, he asks the uncomfortable question: "Why do we find the remains of so many unknown species?" He answers his own question by saying they were all destroyed by ancient cataclysms, "revolutions on the surface of the Earth." It is an unambiguous and unqualified endorsement of extinction. The uncompromising nature of his assertion and his own growing prestige made it a claim to which attention needed to be paid.

The rules of the institute did not permit Cuvier to make any changes in his paper after it was read at an open meeting. In the months that passed between his reading and the publication of the next edition of their journal, Cuvier acquired additional relevant and important information. His first burst of short papers had made him something of a scientific celebrity. Savants all over the continent began corresponding with him, sending notes of their discoveries and suggesting collaborations. During that short time, he had received a copy of Blumenbach's *Handbuch* and notification of no fewer than four other unknown species related to elephants. One (or possibly two) of those species he placed into the genus *Elephas*. These were represented by teeth found in Europe and South America. The European teeth were the source of the pseudo-turquoise found near Simore that Réaumur had investigated almost a century earlier. The South America teeth had been sent to him by the German naturalist Alexander von Humboldt from Venezuela. He placed two more species in the pachyderm family, but in a different, not yet determined, genus than *Elephas*. The final unknown, he said, belongs somewhere between the pachyderm family and other ruminants. He reported this news in an addendum to his paper and promised a more detailed paper at a later date.

The evidence of extinct species continued to pile up. In 1801, the institute published a plea to the international community, written by Cuvier, for a major effort to discover and identify extinct species. In the five short years since he suggested that the mammoth and the Ohio animal might be extinct, he had identified twenty-three species that he proclaimed were definitely extinct. Along with the three elephant species, he now added the German cave bear, the Irish elk, the Siberian rhinoceros, the megatherium, some large fossil turtles, a large crocodile-like animal discovered in the Netherlands, and a small flying lizard that he named pterodactyle. For his part, he planned to start his search for extinct species in the region immediately surrounding Paris. This had three distinct advantages: it would have instant appeal to potential audiences, it was relatively unworked terrain, and most of the heavy labor could be done in gypsum quarries. Over the next ten years, he published a regular stream of papers in which he identified more extinct species and entire genera. In 1812, he gathered them all together in a volume titled *Recherches sur les ossemens fossiles de quadrupèdes* (Researches into the bones of fossil quadrupeds). In an article written in 1806, and included in the collection, he identified five species with the same cusped teeth as the Ohio animal. He had now seen enough bones to feel safe in creating a separate genre for these fossils. He called it *Mastodonte*. The conical cusps reminded him of a woman's breasts, so he created a compound word using two Greek roots: *mastos*, meaning breast, and *dont*, meaning tooth. In a letter to Charles Willson Peale, Thomas Jefferson translated it as "bubby toothed."

Cuvier was by no means the first to claim an animal was extinct, but he was the most assertive. Beginning in the 1760s, Collinson, d'Holbach, and others had suggested it for mammoths and mastodons. Bumenbach's "unknown" category left open the possibility that certain fossils might represent species that no longer exist. But Cuvier's lists raised extinction from the realm of anecdotes about one species or another as isolated incidents into the realm of a principle of nature. From time to time, entire assemblages of life vanished and were replaced by new ones. His catastrophes or revolutions accounted for the disappearances. He was never clear about how the new species came into being. He did not believe in multiple divine creations nor did he accept transformationist (evolutionary) solutions. It just happened. Cuvier's argument for extinction began with the *Megatherium* and the missing elephants. There was not enough room in the world for such large, unusual animals to go unnoticed. But there was so much more to know about them than that they were gone. His own methods of comparative anatomy made it possible to deduce some facts, such as diet and locomotion. More than bones would be needed to tell the rest of the story.

CHAPTER 8

THE FIRST GREAT MAMMOTH

The most important mammoth discovery of the eighteenth and nineteenth centuries came about because of a diplomatic failure, two strangers meeting in a frontier town, and an Evenk tribesman beachcombing during the short break between the fishing season and the reindeer hunting season. Great discoveries are frequently the result of hard work, of someone with the right training who knows what they are looking for and who looks in the right place. Other discoveries are the result of plain dumb luck, of a singular course of events that puts the right person in the right place at the right time. The recovery of what would eventually become know as the Adams mammoth is one of the latter. But, we're getting ahead of ourselves. Let's take things in the order they that happened.

Ossip Shumachov was a chief of the Batouline clan of the Evenki, a Tungusic people who occupied a huge but sparsely populated swath of Siberia from the Yenissei River to the Pacific and from Manchuria to the Arctic

Ocean. They shared the center of this territory with the Yakuts (Sakha), from whom the Evenki of the south learned horse herding. Above them lived the forest Evenki, nomadic hunters and trappers who lived a solitary life in the forests, rarely gathering in groups larger than an extended family. The Batouline lived in the far north. The land they called home was productive enough that they lived in cabins in a small village called Kumak Surka, the northernmost permanent settlement on the Lena River. Above them, there was nothing but seasonal hunting and fishing camps. They hunted wild reindeer, fished the river, and owned domesticated reindeer. Like all people of the north, they trapped during the winter in order to have a commodity to trade with visiting Russian merchants in exchange for metal tools, powder and shot for their muskets, tobacco, and luxury items such as colored fabrics. Living on riverbanks and near the ocean shore as they did, they also hunted for ivory.

In the summer of 1799, after the main salmon run in the Lena had ended, Shumachov and most of the Batouline moved to the Bykovski Peninsula on the coast of the Arctic Ocean. They frequently visited the peninsula during the summer, as it was the calving ground for their reindeer herd. At the end of the summer, this same peninsula was also where they conducted the wild reindeer hunt. The hunt involved driving the migrating, wild reindeer into fenced areas where they could easily be killed. After building the teepee-like huts that served as their temporary homes, the Batouline repaired the fences and waited for the herds to pass by on their way to winter pastures on the Lena Delta. While on the peninsula, they ate fresh local foods, berries, and saltwater fish, rather than the dried fish that they had stored for winter. The saltwater fish they caught were considered something of a treat. If they had some spare time, while waiting for the reindeer, Shumachov and his brothers scoured the beaches for ivory or anything else of value that might have washed up.

It was probably during a fishing trip that Shumachov made his discovery. He was paddling his kayak-like canoe along the seaward side of the peninsula, below a hill called Kembisagashaeta, when he noticed a dark mass just beginning to be exposed near the top of the frozen bluff. He went ashore and climbed up on a rock to get a better view of it, but could not tell what it was except to say that it was not like the driftwood that he sometimes found embedded in the permafrost near the sea. With more important things to do during the remainder of the short summer, he did not spend any more time investigating the mass.

The following summer, when his family returned to the peninsula, Shumachov found a dead walrus on the beach below the mysterious mass. I am not sure why Adams, the person who recorded Shumachov's story, thought the walrus was important enough to mention. It might be that, in stopping to look at the walrus, Shumachov was reminded of the mysterious mass on the hill and attempted to examine it again. Shumachov's primary reason for investigating a dead walrus would not have been to harvest it for food; a dead walrus on the beach would be a smelly mess in no time, though a welcome buffet to seagulls, foxes, and polar bears. The only practical reasons for giving it a second look would have been to see if the thick skin was salvageable or if it had tusks that he could cut off and sell. Having looked at the walrus, whatever he thought of it, Shumachov also checked the mysterious mass on the bluff. By now, enough had eroded out that he could distinguish that it was made up of two separate large parts, but he still could not tell what they were.

By the third summer, in 1801, Shumachov was able to identify the mass as a mammoth. The two parts he had seen the previous summer were revealed as feet, one flank was visible, and, most importantly, he could see one of the tusks. Returning to the summer camp, he told the others about his discovery. Shumachov expected the news to be cause for celebration; instead, the older members received it with expressions of sadness. The old men explained to him that several generations before, a hunter had discovered a mammoth carcass near the same spot. He and his whole family died soon after. Because of that, the people of the region viewed mammoth carcasses as portents of disaster. Shumachov became sick with worry and retired to his cabin to await his doom. After several days of not dying, he decided the old superstition might be not true and began to think about the value of the tusks, which he described as large and beautiful. He called his brothers and returned to the bluff. They did not have the time or the tools to chip them out of the rock-hard, frozen ground, so they covered the whole carcass with grass and left a guard to keep other people away. The next year was colder than usual and the mammoth did not thaw any further. It is possible that Shumachov's efforts at hiding his treasure insulated it from whatever warmth there was that year.

Finally, at the end of the fifth summer, in 1803, the bluff had eroded and thawed enough for the mammoth to break free and tumble down onto the beach. The following March—in the dead of winter, a time of little activity before the spring hunting and fishing season arrived—Shumachov and two companions left their village and returned to the peninsula to collect the ivory treasure. The tusks were nine feet long and weighed two hundred pounds each, a bit larger than the average set of tusks collected by ivory hunters, but not extraordinary in size. A few days later, Roman Boltunov, a merchant from Yakutsk, arrived in Kumak Surka and bought the tusks. The price was fifty rubles' worth of trade goods—roughly two thousand dollars in modern currency. For a people who lived almost completely outside of the money economy, this would have been a tremendous boon to the village.

When he heard that the ivory came from an intact mammoth, Boltunov asked Shumachov to show him the carcass. As an ivory trader, Boltunov would have known how rare sightings of mammoth carcasses were. He would also have known that educated Russians from the West and other Europeans were quite interested in mammoths and would pay handsomely even for information about them. He decided that it was definitely worth his time, at least four days, to get a firsthand look at the mammoth. It was snowing heavily when they got to the site. Scavengers had already discovered this large block of free meat and eaten parts of it. Much of the face had been torn away. Still, the majority of the body was still there and in one piece. Boltunov cleared away enough snow to get a good look at it and examined the head. What he saw was bigger than any animal he had ever seen or heard of. It was covered with long, rust-colored hair. It had a fat body and thick legs. He made some measurements on the spot. After returning, he wrote down some of the details and later, on the opposite side of the same sheet, made a drawing of it from memory. He was correct that the trip would be worthwhile; when he returned to Yakutsk, the head of the merchant's guild bought his notes and drawing.

At first glance, Boltunov's drawing is laughably wrong; it looks like a mutant combination of a boar and an elephant. But, considering the information he had to work with, it is not a bad reconstruction. It demonstrates an intelligent mind and an active curiosity attempting to extract the most from a small amount of information. It is very possible, even likely, that Boltunov had never seen a picture of an elephant and had no reference point for "elephant-ness." He would, however, have seen a boar. Most large mammals he would have been familiar with—dogs, cattle, horses, and reindeer—had long relatively thin legs and heads that rose up from the body. Only bears and pigs had thick bodies, heavier legs, and heads that protruded forward from the body on almost imperceptible necks. In his written description, he twice refers to the mammoth's "swinishness."

The trunk was gone when he saw the carcass; the bloody base of the trunk could very well have resembled a pig's snout. The tusks in his drawing look bizarre; one seems to be pointing up while the other points down. Karl von Baer, who examined the drawing in the 1860s, believed that Boltunov was inexpertly trying to indicate that he believed the tusks should have pointed outward. Even in Baer's time, most scientists incorrectly believed they pointed outward. Yet Boltunov correctly placed the tusks in the upper jaw, not in the lower as they would have been in that of a boar, and he was actually trying to show that the tusks are pushed *together* in the snout, which is in fact correct. Mammoths' tusks start much closer together than those of living elephants and curve out before curving around and back in. Nevertheless, the eyes in this rendering are far too high on the head, and the drawing also shows tiny ears on top of the mammoth's head, which do not match Boltunov's written description. There, he says the ears were six *vershoks* (about eleven inches) long and on the "outside" of the head. The problems with the eyes and ears are probably the result of faulty memory and the amount of time that had passed between committing his first observations to paper and making the drawing. The body is more elephant-like than boar-like, boxy, with pillar-like legs and a short tail. The only other boar-like details on the body are what appear to be fetlocks and thin hooves. The hooves might be his interpretation of the elephant's toenails as a cleft hoof. Boltunov drew little lines around the mammoth that indicate hair running the full length of its body. Finally, at the top of the page he made a separate drawing of a mammoth's tooth, which, with its washboard surface, would have been very different from those of any mammal he was familiar with.

In the 110 years since Evert Ysbrants Ides first reported the discovery of a mammoth carcass in 1692, only four more had been reported, and none of them had been recovered, made available for European scientists to examine, or even very well described. Dozens more were probably discovered during that time, but not reported. Pallas's half of a woolly rhinoceros, from 1772, was the closest that the eighteenth century had to offer to a mammoth until Shumachov's discovery. But as luck and fate would have it, Mikhail Adams, a naturalist from St. Petersburg (not from Scotland or England, as is sometimes reported), was also in Siberia at this same time. Although only twenty-seven, Adams was already a veteran field biologist. Soon after the kingdom of Georgia was annexed to the Russian Empire, he traveled in the entourage of General Apollo Musin-Pushkin to inspect economic resources of the new territory and brought back samples of several previously unknown species of flowers.

The series of events that brought Adams to Siberia began, in part, when a Russian sailor named Adam Krusenstern found himself with time to spare in Canton, China. Krusenstern was one of a group of young officers who had been sent to spend a few years training with the British Navy. While several of his comrades spent the years fighting the French Navy during Britain's on again, off again war with the revolutionary republic, Krusenstern took every opportunity he could to see the world. By transferring from ship to ship, he went first to the United States and the Caribbean, then to the Cape Colony (South Africa), which the British had just seized from the Dutch East India Company, from there to India, and, finally, on to China. At every step along the way, Krusenstern was impressed with the advantages that would come to Russia if they could increase the amount of trade they conducted by sea. It was in Canton that his ideas for trade came together to form a grand plan.

In 1799, some months before Shumachov spotted his mammoth, Krusenstern watched a small American ship arrive in Canton from the Oregon Country. The ship carried a load of furs that sold quickly and brought top prices. Why, he wondered, couldn't the Russians do the same and bring furs directly from Alaska to sell in Chinese ports? At the time, the only point at which Russia was allowed access to China was the Siberian border town of Kiakhta. The terms of that trade had changed very little since the time of Peter the Great.

Nothing was satisfactory about that arrangement. The trade was regularly shut off at the whim of the Chinese court. Hauling goods overland from European Russia to Kiakhta often damaged them, lowering their value. The shipping costs were high. For Alaskan furs, the situation was simply ridiculous. Furs had to be carried on unsatisfactory ships from Alaska to Okhotsk, on the Siberian coast, overland to Kiakhta, where they were sold to Chinese merchants who carried them across the Gobi Desert to their final destination in cities near the coast. A robust sea trade would not only make commerce with China easier and more profitable; it would solve a host of problems with managing the eastern end of the sprawling Russian empire. Every supply, from bread to nails, that went to the Russian Far East and Alaska had to come from European Russia by pack train, a trip that could take up to a year. Krusenstern envisioned a trade in which supplies for the colonies would be brought from European Russia across the Atlantic and around South America. After delivering the supplies to Alaska and Okhotsk, the ships would take loads of furs directly to their primary market on the Chinese coast. Finally, after selling the furs, the ships would return to Russia carrying Chinese luxury goods.

When he returned to St. Petersburg, Krusenstern wrote a detailed proposal and submitted it to his superiors. He arrived at a bad time. Both the admiralty and commerce ministry were undergoing administrative shakeups and the people he had expected to look favorably on his plan had all been ousted. The new regime appeared uninterested. Convinced his plan would never see the light of day, he considered retiring from the navy.

Krusenstern should not have been so impatient. Two years later, the tsar, Paul I, was assassinated and a group of forward looking ministers was appointed by the new tsar, Alexander. Among them were Nikolai Mordvinov at the admiralty and Nicolai Rumiantsev at the commerce ministry. They discussed Krusenstern's plan, found sponsors, and presented it to the emperor. What Krusenstern did not know was that for years there had been support at the court for some kind of expedition to the Pacific. The ministers knew the advantage of expanding trade with China. They knew how badly Alaska was managed. They felt that it was necessary for the prestige of Russia to send a scientific expedition to the South Seas similar to those that James Cook, George Vancouver, and Jean-François Lapérouse had made for the British and French. Mordvinov and Rumiantsev were both supporters of a Pacific expedition and pushed through a plan in, what was by Russian bureaucratic standards, a record time of barely a year.

Krusenstern had some idea that his plan was being discussed, but it came a surprise to him when, in August 1802, he was notified by the admiralty that an expedition based on his plan had been approved and that he was in charge. Furthermore, they wanted him to find two ships, outfit them, hire a crew, and leave before winter. The plan he was handed was vastly more ambitious than anything he had envisioned and took much longer to pull together than the two months Rumiantsev had initially ordered. By the time he left, a full year later, Krusenstern's commercial voyage to China included a full diplomatic mission to Japan, a major inspection tour of operations in Alaska for the Russian-American Company, scientific projects everywhere on the route, orders to map badly understood parts of the Siberian coast, and anything else anyone with influence at the court could think of.

The mission creep and bloat that had happened to the China project were not limited to the sea voyage. While Krusenstern prepared his ships, Rumiantsev began planning a second approach to China, by land, through Kiakhta. For years, the vice-minister of foreign affairs, Adam Czartoryski, had been developing his own plan to open China to greater trade. With Rumiantsev's support, he was able to create a great embassy with the goal of showing the flag in Beijing and improving the terms of trade at Kiakhta. The embassy was originally planned as consisting of forty people headed by Count Yuri Golovkin. By the time it left St. Petersburg, in July 1805, it had swollen to nearly three hundred people, including younger sons of the very best families, their servants, a fourteen-piece orchestra, Cossacks, and a contingent of scientists, including Mikhail Adams. The final objectives of the mission included opening the entire Chinese border to trade, allowing Russian merchants complete freedom to travel inside China, establishing a permanent legation in Beijing, opening the Amur River to Russian boats, and setting up a trade station in Canton. To make the journey easier, the embassy traveled in four groups that would reunite in Irkutsk. The scientists were assigned to the last group.

At the same time that Adams was traveling east across Siberia with the embassy, Krusenstern and his scientists had already been on the far side of the continent for more than a year. Following a voyage from St. Petersburg, south through the Atlantic, around Cape Horn, and across the South Pacific, they had arrived in St. Peter and Paul on the Kamchatka Peninsula in July 1804, just before Schmachov's mammoth finally broke loose from the bluff and tumbled to the beach. Krusenstern spent the following year exploring and mapping the waters around Japan and eastern Siberia and supporting Nikolai Rezanov's diplomatic mission to Japan. That mission ultimately failed because, among other reasons, the temperamental Rezanov refused to make the deep bow demanded of him by Japanese protocol. Krusenstern and Rezanov did not like each other, and Krusenstern was happy to see Rezanov take one of the expedition's ships and sail off to Alaska, where he would plot the annexation of the west coast of America from Sitka to San Francisco.

Rezanov's failure was his own and did not reflect on Krusenstern. The scientific component of the mission was a resounding success. His team corrected maps, made astronomical observations, collected botanical and zoological samples, and studied the flora, fauna, and people of the Marquesas Islands. Many of the expedition members went on to stellar careers in exploration and the sciences. The mission cartographer, Fabian von Bellingshausen, later discovered Antarctica in 1820. Krusenstern's cabin boy, Otto von Kotzebue, became an important Pacific and Arctic explorer in his own right. The ship's doctor, Grigory Langsdorff, became a diplomat and an important explorer in, of all places, Brazil. However, the scientist most important to our story was Wilhelm Gottlieb Tilesius.

Tilesius was a doctor and naturalist from Mühlhausen in central Germany. At thirty-four, he already had one scientific expedition under his belt, having traveled to Portugal in 1797 with Johann Hoffmannsegg to collect marine samples. Besides those credentials, he was also a first-rate scientific illustrator. This would be one of his primary responsibilities on the Krusenstern expedition. If he had done nothing else in his life, his paintings of the Nukuhiva natives would be enough for him to be remembered.

Krusenstern thought enough of Tilesius that he renamed Mount Iwate in Japan Mount Tilesius. (Not surprisingly, the Japanese preferred to keep the original name.) While on this same expedition, Tilesius provided the first scientific descriptions for dozens of species of marine life, including the impressive and delicious king crab—*Paralithodes camtschatica*.

In August 1805, when the ship returned to Kamchatka for the third and last time, Tilesius heard that the captain of a supply ship that traded on the Arctic had recently seen a mammoth carcass. Captain Patapof claimed to have "lately seen a Mammoth elephant dug up on the shores of the Frozen Ocean, clothed with a hairy skin." As evidence of this claim, he gave Tilesius some hair that he had cut from the animal's hide. Tilesius met Patapof more than a year after Shumachov harvested and sold the tusks from his mammoth. It is possible that the two mammoths were one and the same or, perhaps, that Patapof's mammoth was another that Shumachov said had been seen two years before he found his. However, it's more likely he saw an unrelated one farther east. The journey from the Pacific to the Lena Delta was a difficult one at best and in most years not possible at all. Tilesius would have spent more time interrogating Patapof if the two ships had stayed together longer. As it was, Krusenstern and Patapof were both eager to put back to sea and, for Tilesius, at the time, the mammoth was no more than a novelty. His real interest was in marine biology. He packaged up the hair sample and sent it to Blumenbach at Göttingen, assuming he would be more interested in it.

While Krusenstern and his scientists sailed south to Canton, Golovkin and his scientists finally arrived at Kiakhta on the Chinese border. There, they were delayed for three months. The Chinese authorities were amazed at the size of the embassy. The Beijing court had only made plans for one hundred Russians, not three hundred. This was more than a matter of permission—though that was very important; the size of the embassy posed very real practical problems. Caravan stops in the Gobi Desert needed to be stocked with food and remounts for the Russians. The border authorities sent to Beijing for instructions and began negotiating with Golovkin. By the first of January, the Russian party had been pared down to 128 essential members. Even then, this number included many of the young dilettantes, the orchestra, and, thankfully, the scientists.

After twenty days of travel in the dead of winter, their caravan reached Urga (modern Ulan Bator, Mongolia), where they were ordered to pause. Word had come from Beijing directing the local governor, the Jiaqing emperor's brother-in-law, to test Golovkin's willingness to observe proper court etiquette. Most importantly, they wanted him to perform the kowtow, bowing to ground and touching his forehead there. The request was sprung on him by surprise. The entire embassy was invited to a formal outdoor reception in the morning. The temperature was –35°C, but everyone still showed up in full formal attire. Golovkin was shown a gold screen representing the emperor's throne and told to perform the kowtow. Like Rezanov in Japan, he indignantly refused. For two hours, he ranted while the rest of the Russians retired to the warmth of their yurts to eat breakfast and eavesdrop. Golovkin said he would only kowtow to the emperor. Apparently, the Chinese authorities did not trust him to do that. Or, maybe they just wanted him to go away. On February 10, 1806, orders arrived from Beijing telling him to leave China. That same day, orders arrived in Canton instructing the local authorities not to allow Krusenstern to trade there, possibly as additional reprimand against the Russian delegation. For Krusenstern, they were too late. He had already sold a load of Alaskan furs and left port the previous day.

By diplomatic and commercial standards, both missions were failures. But by scientific standards, both were successes. When Krusenstern left Canton, he sailed directly back to Europe by way of South Africa. His scientists had few opportunities for further research, but what they had already done in the South Pacific, Kamchatka, and Japan would fill volumes. Tilesius published a volume of drawings of the flora, fauna, and people of the Pacific that made important contributions to the understanding of all three categories. Bellingshausen published an atlas of the newly explored coasts and islands they had visited. Rezanov studied the Japanese and Polynesian languages.

When Golovkin was kicked out of Urga, the opportunities for his scientists were just beginning. The embassy disbanded in Irkutsk that March with Golovkin, the dilettantes, and the orchestra heading back to St. Petersburg; the scientists were allowed to stay behind. The credentials given to them by the academy allowed them to work independently of the diplomatic mission. Joseph Rehmann collected medicinal herbs along the Mongolian border. Lorenz Pansner studied mountains in Central Asia. Julius Klaproth, who wrote one of the only accounts of the Golovkin mission, studied the Mongolian, Tibetan, and Buryat languages. One of the questions he asked his new acquaintances was the origin of the word "mammoth" and what they believed about the animal. One story he recorded was that mammoths had been on the ark with Noah. After the Flood, they had tried to move north, but found the waterlogged earth too soft, and sank to their deaths in the mud, which later froze, preserving them. Ivan Redovsky and Mikhail Adams both went north to Yakutsk, hoping to collect plant specimens in the summer. Redovsky traveled east across the Aldan Mountains toward Kamchatka. Adams planned to travel down the Lena River to its delta.

At some point, after arriving in Yakutsk, Adams met a Mr. Popoff, the head of the merchant's guild. Popoff must have known that Adams was planning on following the Lena to its delta. When they met, he brought Boltunov's drawing and description and told him "that they had discovered, upon the shores of the Frozen-Sea, near the mouth of the river Lena, an animal of an extraordinary size: the flesh skin, and hair, were in good preservation, and it was supposed that the fossile production, known by the name of Mammoth-horns (mamontovikost), must have belonged to some animal of this kind." Adams dismissed the drawing as "very incorrect," but forwarded it to the academy in St. Petersburg along with a letter explaining his plans to put aside his plant collecting and rush to the delta to see what, if anything, of the mammoth could be salvaged at this late date.

It took Adams a few days to prepare before he could leave Yakutsk. He needed to secure transportation and supplies. He needed to get his papers in order. He needed to inform the local authorities of his change of plans and convince them that he had permission to do so. He also needed certain letters. The English translation of his account uses the genteel phrase "indispensable letters of introduction." In fact, these letters would have been no-nonsense orders that gave him the authority to command the assistance of imperial officers and civilians, and to requisition supplies, housing, transportation, and labor with nothing more than a vague promise of later payment. Without these letters, he would only have been able to recover whatever he could personally carry. Where mammoths are concerned, that's not much. The journey of a thousand miles from Yakutsk to Kumak Surka took three weeks, and Adams reached his destination at the end of June. Bad weather kept him there until August. Adams spent the time studying the people and admiring the landscape. Though he doesn't mention it in his own account, he must have spent some of the time collecting botanical samples. Another member of the failed China embassy wrote of "a large number of plants, an herbarium sibiricum" from him arriving in Irkutsk during the summer to be forwarded to the academy.

It was August before Adams could travel to the Bykovsky Peninsula to look at the mammoth site. To follow the Lena to its delta and then follow the coast to the peninsula was a journey well over a hundred miles. Fortunately, there were several overland shortcuts. Shumachov took what was called the deer trail, a straight track across Kharaulakh Ridge, a northern extension of the Verkhoyansk Mountains, then down the Khorogor River, which leads directly to the base of the peninsula. This was not the easiest route, but it was the most direct. The party consisted of Adams; three Cossacks who had come with him from Yakutsk; a merchant named Belkoff, who had carried them down river from the town of Schigansk; Shumachov, leading the way; and ten men from Kumak Surka. They made camp on the hill Kembisagashaeta, directly above the place where the remains of the mammoth lay. From there, Shumachov's help finding the mammoth was a mere formality; it could be smelled a mile away.

Adams's first sight of the mammoth was not encouraging. During the two years since Shumachov had removed the tusks, the carcass had been at the mercy of local scavengers. Most of the flesh and organs were gone, along with the trunk, the tail, and one of the forelegs with the same shoulder blade. Sakha herders, who shared the peninsula with Shumachov's people, had discovered the carcass and fed its meat to their dogs. But further inspection showed it to be a scientific treasure. Two of the feet were completely intact, with skin and flesh still covering the bones. One eye and the brain had dried up, but were still in place and undamaged by predators. Most of the skeleton was intact, and many of the bones were still held together by ligaments and skin. Even with a leg missing, this was by far the most complete mammoth skeleton ever recovered.

Adams wasted no time getting the mammoth ready for transport. He made some measurements, then pulled the skin back and disarticulated the skeleton into pieces that could be moved. Next, he had the skin lifted in one piece. It was "of such an extraordinary weight, that ten persons . . . moved it with great difficulty." He had those ten men spread it out on driftwood to dry. The side of the head facing the ground included the bonus of an intact ear. He reported that the skin was "of a deep grey, and covered with reddish hair and black bristles." The reddish guard hairs were over two feet in length. Most of the hair had fallen off the skin before Adams arrived and nearly all of the rest fell off when the skin was dried for transport back to St. Petersburg. With the skin drying and the bones ready for transport, Adams dug in the ground underneath and around the mammoth. He was rewarded by finding the missing shoulder blade. Adams carefully gathered all the hair he could find, bagging up over forty pounds.

With everything packed up for shipping, and Belkoff not having arrived yet with the boat, Adams took a few days to explore the peninsula and make observations on the local geology and botany. Naturally, he collected plant samples, but he also made some offhand observations of the ground that are the best descriptions of permafrost up to that time and that have been completely ignored ever since. Near one of the small lakes in the peninsula, he examined the cabin of Ivan Bakhov and Nikita Shalaurov, who had stayed there during their unsuccessful attempt to sail from the Lena eastward through the Bering Strait in 1760. He had two giant crosses put up to celebrate the discovery, one for the mammoth and one for himself. At last, Adams tired of waiting for the boat and returned overland to Kumak Surka. The boat with the mammoth remains arrived there a week later.

Before returning to Yakutsk, Adams boiled the bones to remove the ligaments and clingy bits of skin. This could easily have been the first time that mammoth soup had been made in more than ten thousand years. He does not mention trying it. The only parts that he did not boil were the skull and feet. These he left covered with skin, and they have remained that way to this day. They were dessicated enough that they needed no extra preparation. With the rest of the bones clean and shiny, he packed things carefully into the bottom of the boat for the trip back to Yakutsk. The one disappointment Adams had with the mammoth was that he had not been able to get to it while the tusks were still attached. His disappointment turned to joy in Yakutsk, where Popoff let him know that he still had that very set of tusks and that he was willing to sell them to Adams for a reasonable price. It would be years before anyone would figure out that they came from a completely different, smaller mammoth and that Popoff had swindled Adams. At Yakutsk, Adams repacked the mammoth one last time—this time onto sledges—and sent it off to St. Petersburg. He stayed in Yakutsk for a few more weeks to collect more plants.

By way of announcing the discovery of the mammoth, the academy published Boltunov's notes in their Russian-language layman's gazette, *Technological Journal*, in November 1806. By then they knew that Adams had recovered a good part of the skeleton and that it was safely en route to St. Petersburg. They planned to reassemble the skeleton for display. For reasons that are unclear, they chose Tilesius for this task. Perhaps, based on the hair sample that he sent to Blumenbach, they thought he had a special interest in mammoths. If so, they would have been wrong. Perhaps they were impressed with his artistic and analytic talents and thought that those would be an advantage when it came to reassembly. Tilesius had been back in the capital since August, having returned aboard Krusenstern's flagship, the *Nadezhda*. He had an enormous amount of work to do. He needed to properly organize and store the samples he had collected. He needed to write up and publish descriptions of those items he believed were unknown to European science. He needed to prepare his many illustrations for publication in the official report of the expedition. He was a marine biologist. None of this mattered; the academy wanted him for the job. Having been chosen, he performed the task in a conscientious and creative manner, although with a great deal of grumbling.

Ninety percent of the reconstruction was not especially difficult. No one doubted that the mammoth was similar in anatomical structure to an elephant. The museum where Tilesius worked, the Kunstkamera, established by Peter the Great for his personal collection of oddities, had the complete skeleton of an Asian elephant for him to refer to. This elephant had been a gift to Peter the Great from the Shah Husayn I of Persia in 1713. He also had Peter Camper's excellent anatomical study of elephants to give him a crash course in vertebrate anatomy. The missing and most damaged bones were all on the same side of the body, making it easy for the museum's craftsmen to make replacements by making mirror-image copies of the equivalent bones on the other side of the body. The three problems that faced Tilesius and required informed guesswork were the length of mammoth's tail; the curve of its spine, that is, its posture; and the correct positioning of its tusks.

When Boltunov saw the mammoth in March 1804, he saw that it had a short tail, measured it (six *vershoks* or eleven inches), mentioned that fact in his notes, and included it in his drawing. When Adams saw the mammoth, two years later, the tail was gone and he wrote that he did not believe it had had a tail in life. Inexplicably, the English translation of his account changed that to say he did think that it had a short tail. Tilesius, on examining the skeleton, saw that there had been more vertebrae beyond the hips, thus indicating a tail. If the mammoth was similar to a typical Asian elephant, like the one of Peter's that he was using a reference, it would have had twenty-four to thirty-three caudal vertebrae. Tilesius guessed that a mammoth's tail should be shorter than an Asian elephant's and had the craftsmen make eight vertebrae for the reconstruction. The correct number would have been in the high teens.

The line of the back was another educated guess. Both of Tilesius's references—Peter's elephant and Camper's monograph—were based on Asian elephants. This would not have bothered him. Based on the study of their molars, it was well established, by his time, that mammoths were more closely related to Asian elephants that to African ones. Tilesius arranged the vertebrae so that the mammoth's outward silhouette would match the shape of an Asian elephant. The shape of a vertebrate's back comes from two elements: first, the line of the inner part of the vertebrae themselves, the part that protects the spinal nerve and is the main structural support for the body; and second, the length of the spines that protrude from the upper side of each vertebra and provide a surface for muscles to attach to. The length of these spines is different on mammoths and Asian elephants. To give the mammoth the round shape of an Asian elephant, Tilesius had to flatten out the inner line of the vertebrae. The correct line as we understand it today, which creates the familiar high shoulders and sloping back, would not be known until the middle of the century, when

Édouard Lartet discovered a prehistoric drawing of a mammoth on a piece of mammoth's tusk at the La Madeleine rock shelter in France.

The last detail that Tilesius had to guess about was the placement of the tusks. No one had yet recovered a mammoth skull with the tusks still attached; the proper placement of them was a mystery. Naturalists who have had the opportunity to examine intact mammoth tusks had all commented on the fact that the curve of them was completely different than that of African or Asian elephants. They curve widely outward as they descend and then inward as they curve up again, and they are much larger than those of any living species of elephant. In the oldest bull mammoths, the tips of the tusks can even cross. The wisdom of the time was that tusks, in any species, are weapons. To this, he added a new rationale; working with the tusks he had, Tilesius believed the tusks had to curve outward or they would have interfered with the mammoth's vision. With that in mind, he made the same assumption the Boltunov had and mounted the tusks on the wrong sides so that they curved out and back like giant ivory scythes. His mistake would not be corrected by the museum until much later, and scientists would still argue about the details of proper tusk placement well into the twentieth century. For more than a century, no one thought to ask the native ivory hunters how the tusks had been situated in the skull before they broke them out for sale. Because graphic art from the nineteenth century is available in the public domain, art directors for news media and magazines, and even science magazines—which should know better—use scythe-tusked mammoths to illustrate paleontology stories to this day.

As soon as Boltunov's notes were published in November 1806, the scientific world eagerly awaited more information. Adams and Tilesius both tried to meet the demand. During the following year, Tilesius and Adams sent copies of Boltunov's drawing, along with skin and hair samples, to Blumenbach and a half dozen other scientists, mostly German acquaintances of Tilesius's. Tilesius obtained permission from the academy to crack open one of their other two mammoth skulls and make a brain cast for the Munich anatomist Samuel Thomas von Sömmerring. They also sent samples to their colleagues in Moscow, which must have pleased them, because they replied by offering Adams a teaching position. One relevant person left out of their largesse was Cuvier. The War of the Fourth Coalition, which lasted from summer 1806 until the beginning of 1808, made direct contact with France nearly impossible. But Cuvier was not completely left out of this latest mammoth dialogue. Blumenbach and others kept him up to date by forwarding information and sharing samples with him. The brotherhood of science trumped politics.

In August, Adams published an account of his journey to the coast and the recovery of the mammoth in a popular St. Petersburg–based French-language newspaper, *Journal du Nord*. He sent copies of his account to the geologist Karl Karsten in Berlin, to the German scientific journal *Allgemeine Geographische Ephemeriden*, and to Joseph Banks, the president of the Royal Society. The newspaper *Berlinische Nachrichten* was the first to publish a German translation, a third-person summary, from Karsten's copy in their October 29 issue. This version was widely reprinted in German and Swiss newspapers. Banks had an English translation made that was published in the *Philosophical Magazine* for November. This version was widely reprinted in Britain and the United States. Despite having been written in French, the first publication of it in French was a retranslation from Banks's English version published in Switzerland and Belgium at the end of 1808. Because of the Napoleonic Wars, a version did not appear in France proper until 1810, when a gazette of English literature published Banks's English translation. That same year a Dutch translation appeared.

The first mention of Adams's account in the journal of the French academy was in the notes of the November 30, 1807, meeting, a full year after his own account had appeared in St. Petersburg. Cuvier reported to the academy that Karsten had sent a copy of the *Berlinische Nachrichten* summary to him and to Bernard-Germain de Lacépède, who chaired the department of reptiles for the National Museum. Cuvier gave an even briefer summary to the members present. After reminding everyone that the proper name for the American mammoth was mastodon, a word he had coined the previous year, Cuvier let them know that he had confidence that this mammoth would confirm his conclusions that the mammoth was a separate species. Even though the German summary had even fewer anatomical details than Adams's original, one detail excited Cuvier. That was the hair. Fur was a good indication that the mammoth was distinct from living species of elephants. More importantly, Adams mentioned that there were two layers of fur: an outer layer of long, coarse guard hairs and an inner layer of wool. This for him was clear evidence that the mammoth had been native to the north and had not swept there by some geological accident. When he wrote his great work on fossils, *Recherches sur les ossemens fossiles* (1812), Cuvier quoted from Adams and lengthened his references to Adams and Tilesius in subsequent editions.

Adams's account, unfortunately, was not much more than a short travel narrative. Though interesting in its own right, the actual description of the mammoth and the conditions of its preservation lacked important details. Many readers were disappointed that he didn't have more to say about the soil and the place where it had been preserved. Adams devotes far more energy in describing the people and landscape of northern Siberia. He waxed poetic about the sparkling, snow-covered mountains; the happy songs and folk dances of the Evenki; and the beauty of Kumak Surka, which he declared should be subject of a national song for the peoples of the north. He promised to write a more extensive tract about the country and its people at a later date. Those who had hoped for more substantial fare about the mammoth were doomed to disappointment. Adams made no drawings and wrote no description of the mammoth in situ. His description of the skin and bones was short and superficial, though he promised to write a detailed study of the skeleton at a later date. Along with the copies of the memoir that he sent to Germany and Britain, he included a note offering to sell the skeleton in order to fund an expedition across the North Pole to Canada.

Meanwhile, Tilesius patiently continued to work on reassembling the skeleton. When it was complete, Emperor Alexander came to see it and was delighted. He bought it from Adams for eight thousand rubles, a considerable sum of money. Adams took the money and departed for his new job in Moscow, forgetting his plans to write more about the mammoth, the people of Kumak Surka, or his expedition to the pole and leaving Tilesius simmering in his wake. Tilesius spent the next two years writing the paper that Adams should have written. In it he tries to maintain the politeness expected in a formal scientific paper, but, despite calling Adams his "illustrious colleague," his contempt seeps through. He carefully points out every instance of Adams using a wrong term of anatomy, even though, being a marine biologist, he would have just learned them himself. He complains that Adams was happy to share details of the Evenki's lives "but important details, such as the position of the mammoth body in the ice and sand, he omitted." He exaggerates every bit of damage to the skeleton until it sounds as if Adams brought back nothing more than a box of bone chips and a ragged piece of rawhide.

As with the physical reconstruction, Tilesius approached his written report in a thorough and conscientious manner. He familiarized himself with the relevant literature from the previous century and corresponded with experts in other fields. In the first half of the report, he makes a plausible speculation of how the native Siberians could have come to believe there were living mammoths underground. He describes the wide territorial range where mammoth ivory had been discovered and exported. He describes the older theories of how elephant bones could have been transported from the tropics to the Arctic. He lists some of the better-known reports of fossil ivory in ancient and medieval literature, like the description of Theophrastus and several English regional histories. He then directs readers who might want to know about other animals frozen in the Siberian earth to Pallas's account.

Meanwhile, when Julius Klaproth returned from Siberia and Mongolia, he went to the academy's library and continued his study of Eastern languages using the collection of Chinese and Manchurian manuscripts stored there. The word and legends of the mammoth from the east continued to interest him. In the library, he located many of the same sources that the Kanxi emperor had used in his monumental natural history. Tilesius had Klaproth write a section on these legends and included it at the beginning of his report directly following Pallas's conclusions on the etymology of mammoth. Tilesius was a great admirer of Pallas, so, even though he includes Klaproth's etymology, he makes it clear that, in his opinion, Pallas's was the correct one.

In his historical review of mammoth scholarship, Tilesius pays special attention to several recent scientific expeditions, using them as a call to action for the academy to send out more expeditions and to have trained people ready to recover important finds before they are lost. Of these expeditions, Pallas and the Vilui rhinoceros have a special place of honor in Tilesius's estimation. He emphasizes the importance of the rhino three separate times. He also mentions Gavril Sarychev who, while travelling from Kamchatka to Yakutsk in 1787, heard about the intact carcass of a great animal recently uncovered on the Alazea River. Though he and his travelling companion were eager to go to the spot, worsening weather prevented them from doing so. Tilesius gives this as an example of the difficulties in recovery even when the right people are present. He quotes extensively from Gmelin and other travelers on the subject of the frozen ground in Siberia pointing out that there is no mystery to the mammoths being preserved for thousands of years; the mystery is how they became buried in the frozen earth in the first place. It is in this section that he tells the story of his receiving the hair of a mammoth from Captain Patapov in Kamchatka.

In the last part of his historical section, Tilesius presents Adams's account. He says this is necessary because it was only published in obscure newspapers that were unavailable to most people. He edited out what he saw as the unnecessary travelogue parts and heavily annotated the rest. Tilesius's notes include useful scientific information, petty corrections to Adams's understanding of anatomy, and endless carping about the various indignities he has had to suffer. One detail of Adams's account that Tilesius did not comment on was the mammoth's sex. Adams, in passing, says the mammoth was a male. He does not say what part of the mammoth led him to that conclusion. Perhaps it was something to do with the skin, the lower part of the skin.

In the second half of the report, he gets into the nitty-gritty of anatomy. In this, he is on more familiar ground. Tilesius was a marine biologist who had never written about mammals but the basics of writing an anatomical description are the same whether for a king crab, a sea cucumber, or a mammoth. He begins with the general characteristics of all elephant-kind, focusing on the skull. He then moves on to the more specific differences between the two living species of elephants, the mastodon, and the mammoth. The purpose of this review is to establish that the mammoth is a distinct species. Cuvier's and Blumenbach's determinations were recent enough that Tilesius felt it necessary to establish this on his own. He did not accept their arguments as proof. Breaking with Cuvier, he states that all four species belong to a single genus, which he calls *Elephantorum*. Adams had come to this same conclusion.

With this established, Tilesius moves on to giving a detailed description of the skeleton that he had restored. To illustrate his points, he prepared two large prints, one of the complete reassembled skeleton with an inset of one of the femurs, the other of a skull and jaw. Because he decided to leave the skin on the skull, he used of one of the academy's other skulls as a model for the second sheet. In this, he gives his readers more than basic measurements of the large bone and completed skeleton. His final mounting was twenty feet long and eleven feet, three inches tall. Because he pointed the tusks outward, rather than inward, their tips were ten feet apart. The whole thing was held together with small iron bars and stood on a sixteen-foot base. He points out important points of muscle and ligament attachments. These are marked with letters and then discussed in the text. In this way, he proves the animal had a trunk, an important point since none of the witnesses saw one.

Tilesius's frustrations would turn out to not to be limited to Adams. He finished his paper in 1810 and submitted it to the academy, but they sat on it for two years before approving it for publication. Another three years passed before it finally was published. That little business of Napoleon invading the country and burning Moscow to the ground might have had something to do with the second delay. When the academy failed to act on his paper, Tilesius had several sets of his drawings printed and sent them to scientists in Europe and America along with letters detailing some of his conclusions.

During these intervening years, Tilesius made the acquaintance of the American ambassador John Quincy Adams. Adams seems to have genuinely liked Tilesius. His diaries from this period mention visiting him several times. On one visit, Tilesius took the future president to the museum and gave him a tour. At the time, the mammoth was on display, but Adams merely noted its existence, giving it as many words as he did Peter the Great's taxidermied horse. On the occasions that he visited Tilesius in his home, the visits seem to have consisted primarily of Adams patiently listening while Tilesius complained about his job and the state of the sciences in Russia. Science was on the decline, he said. There were only two members in the entire academy that measured up to the standards of men of the previous century, such as Pallas and Gmelin. It was impossible to find decent engravers to prepare his drawings for publication. It was impossible to find good paper for the work. Even if they could find decent paper and even if they could sell all the copies, he did not think the people who bought them would read them. He never acknowledges that Napoleon and two decades of war in Europe might have had an effect on the supplies, personnel, and budget available to him.

With the mammoth project out of the way, Tilesius was able to work on writing up his discoveries during the Krusenstern expedition. A bibliography of his writing shows a flurry of small papers appearing over the next five years, culminating in the volume of his drawings from the South Seas. Three of these papers appeared in the same volume of the academy journal as the mammoth paper. Although his professional life flourished during these years, his personal life was in decline. He had married a much younger woman after returning from the Pacific, and it was not a happy marriage. Eventually, she would run away with another man, leaving him with a young son to take care of. At this point, Tilesius decided he had had enough of Russia. He packed up his things, took his son, and returned to Germany in 1814, where he moved back in with his mother.

When the embryology pioneer Karl von Baer joined the academy in 1834, he developed a side interest in Siberian exploration and sought out members who had known Adams and had been present during the restoration of the skeleton to learn more about it, "but they had heard nothing more special and only said that Adams had embellished his report." Adams should be forgiven his weaknesses as a paleontologist and geologist; these were not his areas of expertise. When he heard about the discovery, he recognized its importance and rushed to recover what he could. Had another year passed, it is unlikely that there would have been enough of the mammoth left to add anything to what the European naturalists already knew. Adams also paid far more attention to the people who actually found the mammoth than most naturalists of the time would have. Adams told the story of the actual discovery in Shumachov's voice. He also included some details of the life of the Evenki, though these were strongly colored by the "noble savage, happy child" of nature ideology popular at the time.

The Siberians did not remember Adams as fondly as he remembered them. When Shumachov learned about the generous reward given to Adams by the emperor, he made the difficult and probably frightening journey to St. Petersburg to protest. The skeleton, he said, was a gift from the Batouline to the emperor. It was not Adams's to sell. According to Tilesius, Adams, safely ensconced in Moscow, laughed off the criticism. Shumachov, no doubt, would have been doubly offended to know that the mammoth would come to be known as the Adams Mammoth and that some later retellings of the story would even call Adams the mammoth's discoverer. Shumachov returned to his people, and there he vanishes from Western history. Adams, on the other hand, continued to be remembered in Evenki history. An example from five hundred miles away and three generations later demonstrates this.

In February 1869, a scientist named Gerhard von Maydell was in the second year of an expedition to northeastern Siberia on behalf of the Russian Geographic Society. While wintering in Nizhne-Kolymsk, well above the Arctic Circle, he received a message from Magistrate Ivaschenko of Vekhoyansk (the coldest place in the northern hemisphere) that some hunters had found a mammoth cadaver not far from Nizhne-Kolymsk. Ivaschenko's message gave a detailed description of the location, near the Alazea River, and named the hunters. As soon as the weather permitted, Maydell headed for the location and tracked down the hunters. Their leader, named Foca, denied knowing anything about mammoths. Maydell pressed him and Foca claimed he had not seen the mammoth himself and could not help Maydell find it. Maydell finally had to quote Ivaschenko's message to "prove" to Foca that he had seen the mammoth. Faced with proof and forty pounds of tobacco, Foca relented and agreed to take Maydell to the spot.

At the time, the imperial government was offering a bounty of up to three hundred rubles, supplies, and even a medal to anyone who reported the remains of a mammoth—a fortune in Siberia at the time. In 1929, V. I. Tolmachoff wrote that to his knowledge, between the time Peter the Great first offered a bounty and the revolution, despite the bounty increasing to one thousand rubles by 1914, only one person ever claimed the bounty. Maydell explained Foca's reluctance this way: "the natives of the area have such a bad memory of Adams's expedition that, where possible, they conceal their discoveries because they are afraid of being forced to work and provide haulage."

Foca's reluctance was not unique. In 1882, Alexander Bunge landed on Moustakh Island to set up a weather station as part of Russia's contribution to the first International Polar Year. Moustakh lies a few miles southeast of the Bykovsky Peninsula, where Shumachov found his mammoth. Seeing that the Russians were on the island to stay, the local headman reluctantly showed Bunge the remains of a frozen mammoth that had been found twenty-five years earlier. The locals sold the ivory to a trader, who told the district magistrate about the mammoth. Like Foca's people, the Moustakh natives so feared being dragooned into unpaid labor that, when the local magistrate came to investigate, they buried the mammoth and told him that they had chopped it up and thrown it into the sea. Over the years, they occasionally dug the mammoth up to feed bits of it to their dogs. By 1882, there was very little of it left for Bunge to examine.

The casual exploitation of native labor is an ugly subtext to most scientific advances on the imperial frontiers, and Russia was no exception. Even when the laborers who dug up the temples and carried the specimens to the coast were paid, we should ask whether they had a choice in that transaction. The important papers that Adams brought gave him the power to conscript the local population for labor. In his account, he describes Shumachov's village as having forty to fifty families, all engaged in putting up food for winter. He then mentions taking "ten Toungouses" with him to excavate the mammoth. If they all came from the village, that would mean he took a very significant portion of the able-bodied adult males and kept them from hunting during the time of year that was most essential to their survival. This is the reality that lies beneath Adams's flattering narrative.

Adams and Tilesius were not the last people to study this particular mammoth. As other mammoths were discovered, like Foca's and two others Maydell had the opportunity to examine, the conservators at the museum made minor changes to the skeleton based on new understandings and discoveries. When the first cave paintings were discovered in France in the second half of the century, they modified the line of the vertebrae to show the now familiar high shoulders and sloping back profile rendered in these ancient eyewitness drawings. They remounted the tusks on the correct sides of the head and later replaced them with tusks of the appropriate size. Today, two hundred years later, the skeleton is still on display at the Zoological Institute of the Russian Academy of Sciences. We have carbon-dated the mammoth as 35,800 years old. We know that he was about forty-five years old when he died. We have teased DNA out of his hair. The Adams mammoth remains an important point of comparison for newer discoveries, and the Adams skeleton itself was a culmination of generations' worth of intellectual evolution, from the dismissal of the concepts of giants and mythological beasts to the acceptance of the difficult ideas of extinction and the great age of the earth. Two centuries after his discovery, the Adams mammoth still teaches us what makes a mammoth a mammoth.

AFTERWORD

Shumachov's mammoth was unquestionably the most important mammoth discovery of the nineteenth century. The assembled skeleton put to rest any remaining doubt that the mammoth was a type of elephant, and established it as a distinct species separate from the elephants of Africa and Asia. The dual-layered hair was accepted by most to mean the mammoth was adapted to a cool climate, though arguments still remained over whether it was truly adapted to an Arctic climate. Cuvier weighed in with an influential "no." His belief was that the north had been temperate when mammoths lived there and that the same cataclysm that drove them extinct had also created modern Arctic conditions. His authority was enough to override Adams and Tilesius, who believed otherwise, and thus erroneously convinced most scientists for the rest of the century.

An important side effect of having so many anatomical questions answered about the mammoth with the Adams skeleton was that much of the debate moved from defining the animal to attempting to understand the environment in which it had lived and explaining its preservation. Many, if not most, naturalists agreed with Cuvier, who did not believe that the climate in the past could have dramatically different from that of the present. Permafrost was a complete mystery, and many did not even believe it was real; it had to be an erroneous observation made by ignorant Siberians. Gmelin noted that the soil near Yakutsk remained frozen during the summer, making it difficult to dig wells. No one followed up on his observation. In 1827, Fyodor Shergin, an employee of the Russian-American Company, made a concerted effort to dig a well through the frozen ground at Yakutsk. He finally broke through after nine years work at a depth of 120 meters. Karl von Baer collected reports of frozen soil and produced a map of the Eurasian permafrost region that coincided with the area where mammoth carcasses were found. Three separate northern expeditions made side trips to the site of Shumachov's discovery during the nineteenth century in order to examine the permafrost there. Today that site is called Mamontovy Khayata, and for twenty years it has been the site of a joint German/Russian permafrost research project. This makes it possibly the best understood piece of permafrost on the planet.

Cuvier wasn't finished with elephants when Adams brought his skin and bones back from Siberia. They continued to show up in his later major works. After Tilesius's paper came out, he gave credit to the importance of his and Adams's work. Although he was enthusiastic about naming extinct species, Cuvier had mixed luck with pachyderms. Blumenbach had several months' precedent over him in naming the mammoth, meaning it would be called *Elephas primigenius* and not *Elephas mammonteus.* His clever name for the mastodon genus also lost out. In the nineteenth century, researchers noticed that Robert Kerr, a Scottish surgeon, had named it *Mammut americanum* in 1792 in an update of Linné's system that he translated to English and expanded using Gmelin's plant descriptions. Kerr doubly insulted Cuvier by later producing a bad translation of the first volume of *Recherches*, which gave the impression that Cuvier's catastrophes were identical with the Deluge instead of local events of a nondivine nature.

The number of species and genera of proliferated throughout the century. Many scientists, as they were now called, promoted Cuvier's two genera to the rank of family and created new genera to fill the space between the families and the individual species. By midcentury, some writers were identifying more than forty species. In 1857, Hugh Falconer tried to make sense of it all by eliminating duplicates and reexamining the criteria for categories. He reduced things to one family with two genera and twenty-eight species. It was a good system and lasted for a short time, but the pace of new discoveries and the lure of naming rights left it in the dust. When Charles Osborn tackled the job of systematizing extinct elephants, they had been promoted one rank above family to become the order Proboscidea. Within that order, he identified five suborders, eight families, forty-four genera, and 362 species. Since then the trend has been to reduce the number species. There are currently about 175 recognized species of proboscideans, ranging from elephants and mammoths to mastodons, to four-tusked Gomphotheres, to Deinotheres with tusks on their chins that hooked downward, to the possum-sized ancestor of them all, *Eritherium*. Mammoths had a rough time in all this. They have been moved into their own genus, reunited with living elephants, moved out into a new genus, renamed, split into multiple species, and recombined. Today, there are about ten recognized species of mammoths sharing one genus. Most of these represent different stages of mammoth evolution, as they moved out of Africa and adapted to cooler climates.

Understanding the world of the mammoth has been slower still. A great, but confusing, step was made in 1837 when Jean Louis Rodolphe Agassiz, a Swiss naturalist with a strong interest in freshwater fish, presented a paper arguing that a large part of the northern hemisphere had been covered by glaciers in the not so distant past. The idea was outrageous by the accepted norms at the time, but the evidence, such as glacial moraines hundreds of miles away from the nearest mountains, was so overwhelming and it was accepted in a relatively short time. This was another piece in understanding the world of the mammoth, though at the time they were unsure whether mammoths had died out during the ice age as they unsuccessfully tried to adapt to the coldest stage or if they had died out afterward because they failed to adapt to the warmer world.

In 1864, Édouard Lartet made a discovery that would make it possible to give a relative date to mammoths. Lartet was a wealthy lawyer who spent his summers engaging in his passion for paleontology. Every year he brought back a nice collection of fossils and made several significant discoveries. That summer, he and his English partner, Henry Christy, were working at the la Madeleine rock shelter where archaic humans had once lived. In May, Lartet was showing Hugh Falconer around the site when the workmen came across some artifacts. One was a piece of ivory with scratches on the outer surface. Falconer recognized the image of a running mammoth. Here, at last, was proof that humans and mammoths had lived together. Mammoths were not from some distant age before man. In other caves, painted images and carvings of mammoths were discovered. Of the many hundreds of painted images that have been found in European caves, mammoths are the third most common animal represented.

In 1822, Henri-Marie Ducrotay de Blainville coined the word *paléontologie* to describe the study of fossils and the methods of Cuvier. Besides being a practicing anatomist and paleontologist, he was one of the first to write about the history of paleontology. Thirteen years after coining the word, he wrote a long paper about the Theutobochus controversy more than two hundred years earlier. Blainville located the various pamphlets that had been published at the time as well as other relevant documents and letters. His account is quite hostile to Habicot; Blainville felt his defense of giants marked him as an obstinate and superstitious old man. The occasion of his writing the article was that he believed he had located the bones in a theater in Bordeaux. He was wrong. The last decendent of Marquis Langon wrote a few months later to say that the bones were still in the family and sent some of them to the museum in Paris.

A strong case can be made for the role of proboscideans in making us who we are. Modern elephants reuse the same routes in their travels for generation after generation. In some countries, like Thailand, human engineers have found the elephant roads so well sited that they simply paved them over to make human roads. When our hominid ancestors left Africa, they might have literally followed the elephants. When we reached new lands, how did we know what to eat? The paleontologist Elizabeth Vrba has suggested that our ancestors watched what the local proboscideans ate. In areas like northern Asia and the Americas, which we colonized in an amazingly short time, observing what a familiar animal ate and eating the same would have been a very handy way of measuring the safety of strange plants. If our ancestors would have dug the same roots and picked the same fruits that the local proboscideans did, they would have avoided some potentially dangerous trial-and-error experimentation and discovered avocados.

We journeyed with mammoths and their kin for millennia. They helped us find our way and food to eat, and they also became food. They provided us with the materials to build shelters, make tools, and practice art. Even upon their extinction, they helped us discover the antiquity of the world. Mammoths were a focusing problem for a scientific revolution. Beginning as giants, unicorns, or proof of the Flood, they became a symbol of how strange and mysterious the past had been. Were these animals elephants? If so, what were so many of them doing in the far north and how did they live there? If not, then just what were they and what happened to them? The answers required new tools and new ways of looking at nature and the past. Today, they help us understand genetics and ancient environments. Our enduring fascination with mammoths and the availability of frozen cadavers makes them the best-understood extinct animal. While we have mapped the genome of several extinct species, that of the mammoth has been more closely studied than any other. We now know that they came in several colors, including blond and ginger as well as the familiar chestnut used by most artists. We've discovered a mutation in their blood hemoglobin that improved oxygen delivery in low temperatures. Probing the guts of frozen mammoths, we have reconstructed the flora that fed an entire food chain. Mammoths were the keystone species in the Siberian and Alaskan environments that humans passed through as they entered the Americas—and had possibly been there for thousands of years before that. When the first Americans passed the ice that covered Canada, we know that they found more mammoths on the other side. Beyond them, they found other proboscideans almost to the end of South America.

We followed mammoths. We learned from them. We learned about them and created a new science. We miss them so much that we want to resurrect them from extinction more than any other animal. They are an inextricable part of the history of the peoples of the north and of the New World. They still have much to teach us.

Our journey together continues.

ACKNOWLEDGMENTS

For me, the problem with writing acknowledgments to a first book is that I don't know if there will be a second one. To be on the safe side, I feel like I should thank everyone who has ever inspired me, informed me, supported me, encouraged me, nudged me in the right direction, spoken a kind word, bought me a drink, or been my cat. And, having filled half of this volume with such a list, I would spend the rest of my life terrified that I had slighted someone deserving. So, here's the abbreviated list of people who most affected this book. Without question, first position goes to the women in my life: to my sisters, Ellen McKay and Carol Bohm, who have always been there to catch their little brother when he falls; and to my ex-wife, Tessa Murphy, who patiently listened to years of exposition about mammoths, the scientific revolution, and weird ideas involving mammoths. Next, I'd like to thank the diverse community of science bloggers, commenters, social media friends, and scholars who, over the last nine years, have encouraged me and helped me clarify my ideas and direction. I would also like to thank librarians around the world and the internet itself.

There literally is no way I could have written this book without the internet. This book grew out of a single blog post. I have a great love of conspiracy theories and fringe ideas. Many years ago, I noticed that lost history theories—Atlantis, polar shift, flood geology, and rogue planets—all used frozen mammoths as proof positive of their ideas. I planned a single post establishing a chronology of what was known about mammoths and when. At some point, while looking up the context of a quote, I realized I was going to look all of them up, regardless of the language. I wanted to know where it started. But, the books and documents I needed were in rare book collections all over Europe. That's when the internet saved me.

World War II demonstrated the myopic view of Westerners toward the vulnerability of historical knowledge. Hitler held a special grudge against Serbia for Germany's defeat in World War I. In the blitz against Yugoslavia, the national museum in Belgrade was specifically targeted to erase the Serbian nation. Less than three years later, Allied bombers burned down most of Central Europe. Vast numbers of historical documents were lost. After the war, libraries in Europe began copying rare books and sending microfiches and microfilms to other libraries around the world. Later, this evolved into scanning them, giving access to anyone, anywhere who had access to an internet connection. The collections I have used most are the Bibliothèque nationale de France, the Deutsche Nationalbibliothek, Internet Archive, Biodiversity Heritage Library, the Hathi Trust, the Rhino Research Center, Google Books, and the Guttenberg Project. There are many others.

Last, and most importantly of all, I want to dedicate this book to my parents, Leonard and Mary Jane McKay. They taught me to love words and to see the history in all things.

NOTES

INTRODUCTION

p. ix "I staid here"—Darwin, pp. 146–147.

CHAPTER 1

p. 1 Discovery—Desfontaines, pp. 224–228.
p. 1 Cabinets of curiosities—Findlen, pp. 1–11.
p. 1 Letter of Lesdiguieres—Ginsburg, p. 196.
p. 2 Inventory—Three different inventories have been published. The earliest is in Desfontaines, pp. 224–228.
p. 2 Fragility of skulls—Ginsburg, p. 37.
p. 2 Confirmation by universities—Habicot, pp. 71–72.
p. 3 Contract—Desfontaines, p. 221.
p. 4 Tissot's pamphlet—Translation in Cumston. Note: Cumston's commentary is plagiarized from Fournier. Blainville (*Mémoire*) says that Tissot was not a Jesuit, but gives no reason for this claim.
p. 4 Theutobochus—There are many spellings of his name in historical literature. He would have pronounced it something like Teutobod.
p. 4 Theutobochus against Rome—This story in told in most histories of the Roman Republic. I used Mommsen, vol. 3, pp. 178–191.

p. 4 Grote Mandrenke and coastal change along the North Sea—Behre.
p. 7 Imagination run wild—Ginsburg, pp. 205–206.
p. 8 Giant soldiers—Fournier, p. 244.
p. 8 Request for more bones and bricks—Desfontaines, pp. 219–220.
p. 8 Touring the North—Richer and Renaudot.
p. 8 Habicot—Cohen, pp. 31–33.
p. 10 Lucretius—Céard, p. 37.
p. 11 Augustine—Cohen, pp. 23–26.
p. 11 Classical accounts of discoveries—Mayor, *The First*, appendix 2.
p. 12 Boccaccio—Kircher, pp. 57–58.
p. 12 Abel on cyclops—Abel, *Die Tiere*, pp. 31–33.
p. 13 Species of dwarf elephants—Poulakakis et al.
p. 13 Lucern giant—Plateri, pp. 548–549.
p. 15 Kircher giant comparison—Kircher, pp. 59–60.
p. 15 Zwinger—Céard, p. 58.
p. 15 Maggi—Ibid., pp. 61–63.
p. 15 Gorp—Ibid., pp. 64–66.
p. 17 Riolan's response—Ginsburg, pp. 188–189.
p. 17 Number of bones—This is a more complicated question than most nonbiologists might think. Americans are taught that the human body has 206 bones. Sure, some people might be born with an extra finger or lacking a rib, but 206 is normal, average. This is not what Riolan and Habicot debated. They were arguing over the very definition of "bone." Several parts of the human skeleton that we call single bones in adults begin as multiple, flexible bone pieces that gradually fuse together as we stop growing. Riolan was particularly argumentative about how many bones the sternum (breastbone) qualified as. Riolan, *Gigantomachie*, pp. 10–14.
p. 19 Politics of French medicine—Pasquier, chaps. 31 and 32.
p. 19 Meaning the word fossil—Rudwick, *The Meaning*, pp. 1–3.
p. 20 Riolan on unicorn—Riolan, *Gigantologie*, pp. 46–47.
p. 20 Ctesias—Shepherd, pp. 26–29.
p. 21 Megasthenes—Beer, p. 14.
p. 22 Physiologus—Ibid., pp. 44–71.
p. 24 "effectuall cure"—Ibid., p. 184.
p. 24 Purity tests—Shepherd, pp. 116–119.
p. 24 Marini—Shepherd, pp. 158–161. His marine animal could mean narwhal tooth or walrus ivory. See chapter 2.
p. 25 Marini—Shepherd, pp. 158–161.
p. 25 Bacci—Ibid., pp. 161–163.
p. 26 *Discours* and *Reponse*—Ginsburg, pp. 190–191.
p. 27 Riolan—Cohen, pp. 33–37.
p. 28 *L'imposture*—Morand, pp. 103–105.
p. 29 Letters—Habicot, pp. 70–73.

p. 29 Gilles—Gillio.
p. 31 Peiresc and d'Arcos—Godard, pp. 68–69.
p. 31 The elephant of Claude de Lorraine—Gandilhon.
p. 31 "I was curious enough"—Godard, p. 69.
p. 32 Nivolet to Peiresc—Tamizey, "Un Document"
p. 32 Peiresc on the coin and the tomb—Ginsburg, pp. 187–188.
p. 33 The last tooth—Ibid., p. 213.
p. 33 Krems discovery—Merian, p. 934.
p. 34 Able and Angerer identify the tooth—Angerer.
p. 35 Theophrastus—Mayor, *The First,* p. 274.
p. 35 Quedlinburg—Guericke, p. 155.
p. 36 Mäyern illustration—Valentini, p. 481.
p. 36 About Mäyern—Krafft, p. 202.
p. 37 Protogaea—*Leibnitz*, xxxvii–xlii.
p. 37 Seeländer—Krafft, pp. 178–9.
p. 38 Early ID as rhinoceros—Gottfried Adrian Müller, pp. 345–346.
p. 38 Fraas—Fraas, pp. 38–40.
p. 38 Abel reconstruction—Cohen, p. 45.
p. 39 Dublin elephant—A.M. [Allen Mullen], pp. 3–42.

CHAPTER 2

p. 42 Ohthere's voyage—Hofstra and Samplonius, p. 239.
p. 42 "Chiefly he went thither"—Bosworth, pp. 9–10.
p. 43 Olaf, Karli, and Thorir—Hofstra and Samplonius, pp. 240–241.
p. 44 Williamson—referenced in Walker, pp. 54–55.
p. 44 al-Muqaddasi—Christian, p. 7.
p. 44 al-Biruni—Laufer, "Arabic and Chinese trade," p. 316.
p. 45 Even imported into India—Laufer, *Ivory in China*, pp. 49–50.
p. 45 al-Gharnati—Ibid., p. 32.
p. 45 Old Chinese sources—Ibid., pp. 24–25.
p. 46 Kangxi & Tulishen—Ibid., pp. 26–28.
p. 47 Sushen—Ibid., pp. 52–53.
p. 48 Appearence of *ku-tu*—Ibid., pp. 53–56.
p. 48 Different types of ivory mixed—Ibid., pp. 35–37.
p. 49 Novgorodian expasion—Armstrong, *Russian Settlement*, pp. 10–15.
p. 49 Mansi etymology—Heaney, p. 109.
p. 50 Digges's failure—*The Dictionary of National Biography*, vol. 15, p. 69.
p. 50 James—Heaney, p. 103.
p. 50 Logan—Logan, p. III, iii, 546.
p. 51 Finch—Finch., p. III, iii, 538.
p. 51 Earliest known record—Helimski, pp. 353.
p. 52 Desceliers 1550 map—Van Duzer, *The World for a King,* pp. 47–59.
p. 52 The elephant—Ibid. pp. 82, 165, 185, nn. 270–273.

p. 52 Maps, animals, monsters—George, pp. 21 ff.
p. 52 1553 map—The original was destroyed in a fire; only photographs taken before 1915 survive.
p. 53 Dieppe school—Toulouse, pp. 1550–1554, and Van Duzer, *World*, pp. 13–25.
p. 53 1546 *Dauphine Map*—Currently in the John Rylands Library at the University of Manchester.
p. 53 1547 *Vallard Atlas*—Currently at the Huntington Library in Los Angeles.
p. 53 Waldseemüller 1516 *Carta Marina*—Seaver, "'A very common and usuall trade'" p. 13, and Duzer, *Carta Marina*.
p. 54 Albert's description of walrus hunting—Borgnet, pp. 515–517.
p. 55 Magnus 1539 *Carta Marina*—Only two copies have survived. One is at Uppsala University in Sweden and the other in the Bavarian State Library.
p. 55 Magnus on the walrus—Magnus, pp. 757–758.
p. 56 Herberstein—Herberstein, p. 112.

CHAPTER 3

p. 59 Witsen in Moscow—Peters, pp. 37–45.
p. 61 Importance of the fur trade in the sixteenth century—Fisher, chap. 2.
p. 62 Rise of the Stroganovs—Lincoln, pp. 33–40.
p. 63 Yermak—Ibid., pp. 41–47, and Bobrick, pp. 41–45.
p. 63 Peak of the fur trade—Fisher, pp. 96–103.
p. 64 Witsen discovers Mammot ivory—the longest account is in Witsen, *Noord en Oost* (1785), pp. 742–748.
p. 64 Mention in shipbuilding books—Witsen, *Aeloude*, p. 4, and Witsen, *Architectura*, p. 4.
p. 65 Protogaea—Leibniz, *Protogaea* p. 99.
p. 65 Tentzel—see chapter 5.
p. 67 Sophia during the rebellion of 1682—Hughes, pp. 52–88.
p. 68 Struggle for the Amur—Lincoln, pp. 64–71, and Bobrick, pp. 79–88.
p. 69 The emperor strikes back and the treaty of Nerchinsk—Bobrick, pp. 88–95.
p. 69 Tolerating plagiarism—Peters, pp. 180–181.
p. 69 Dedicated to Peter—Keuning, p. 98.
p. 70 Winius as a map source—Ibid., p. 99.
p. 71 Letter of thanks—Ibid., p. 103.
p. 71 Advice on China and Persia—Peters, pp. 113–116.
p. 72 1692 mammoth—Witsen, *Noord en Oost* (1692), pp. 472–473.
p. 73 Burnet—Gould, pp. 30–41.
p. 74 Ides biography—Hundt, pp. 1–8.
p. 75 Ides mammoth—Ides, pp. 25–27.
p. 77 Ludolf background—Simmons, pp. 104–107.
p. 78 Ludolf mammoth—Ludolfi, p. 92.
p. 79 Hooke—Hooke, p. 438.
p. 80 Peter in Amsterdam—Peters, pp. 124–127.

p. 80 Letter to Cuper—Gebhard, pp. 296–299.
p. 82 Avril mammoth—Avril, pp. 175–178.
p. 82 Avril deported—Hughes, pp. 208–209.

CHAPTER 4

p. 86 Treatment of prisoners—Massie, pp. 638–640.
p. 86 Ivory carvers—Lange, p. 12.
p. 86 Corruption of Gargarin—Massie, p. 930.
p. 87 Müller's report—Johann Bernard Müller, pp. 50–52.
p. 89 Society meeting and drawing—Nordenskiold.
p. 90 Might be a Siödiur—Hag, p. 56.
p. 91 Lost tribes—Rudbeck.
p. 91 Sparwenfeld's analysis—Hag, p. 62.
p. 92 Probably Capt. Tabbert—Hag, p. 63.
p. 93 Strahlenberg background—Ehrensvärd, p. 23.
p. 93 Messerschmidt—Vermeulen, chap. 3.
p. 93 Strahlenberg at the society—Hag, p. 64.
p. 94 Bell on Strahlenberg—John Bell, vol. 1, pp. 191–192.
p. 94 Strahlenberg's map—Ehrensvärd, pp. 24–25.
p. 94 Tatar Chronicle—Ehrensvärd, p. 26.
p. 94 Mamatowa Kost—Strahlenberg, pp. 402–406.
p. 96 Remezov—Bagrow, "Semyon Remezov."
p. 97 Background to Tatishchev's trip to Sweden—Dunér, pp. 933–936.
p. 97 Tatishchev's paper—Tatischow.
p. 99 Linné—Hag, pp. 70–71.

CHAPTER 5

p. 101 Tonna discovery—There all several accounts of the discovery. All agree closely. I have used the version Tentzelii "Epistola de sceleto" published in *Philosophical Transactions.*
p. 102 Submit briefs—Collett, p. 168.
p. 102 Doctors' brief—Anon. [Schnetter].
p. 103 Tentzel's public presentation—Tentzel, *Monatliche*, April 1696, pp. 297–498.
p. 104 Two hundred years old—Ibid., p. 309.
p. 104 "The use of ivory"—Tentzel, "Epistola de sceleto," n.p.
p. 105 May letter from Leibniz—Leibniz, *Sämtliche*, vol. 12, p. 602.
p. 106 Appeal to Magliabechi—Tentzel, "Epistola de sceleto," n.p.
p. 107 Boccone traveled to Gotha—Collet. p. 176.
p. 108 Additional correspondence with Leibniz—Leibniz, *Sämtliche*, vol. 12, pp. 618, 639, 653, 661, 707, and 748.
p. 108 Assures Leibniz that he does not mean extinction—Leibniz, ibid., vol. 13, p. 204.
p. 108 *Scala naturae*—Wilkins, pp. 50–53.
p. 110 Steno and glossopetrae—Cutler, pp. 53–62.

p. 110 Steno's geology—Ibid., pp. 93–122.
p. 111 Early societies and journals—Rappaport, Chap. 1.
p. 111 Origin of the Royal Society—Ibid., pp. 20–26.
p. 112 Mentions Witsen and Ludolf's mammoths—Tentzel, *Monatliche*, January 1697, pp. 52–53.
p. 114 Disappointed with the society's response—Collet, pp. 176–177, n. 530.
p. 114 Cannstatt discovery—Spleiss, n.p.
p. 115 Prodigious Os Frontis—Molyneux, "Part of 2 Letters."
p. 115 Molyneux antlers—Molyneux, "A Discourse Concerning."
p. 116 Nevile and Molyneux elephant—Nevile, "A Letter," and Molyneux, "Remarks."
p. 117 Sloan's first paper—Sloane, "An Account."
p. 118 Sloan's second paper—Sloane, "Of Fossile Teeth."
p. 120 De Bruyn, Schober, Lange & Bell—Vermeulen, p. 116.
p. 121 Breyne recommends Messerschmidt—Ibid.
p. 121 Messerschmidt heartbroken—Messerschmidt, p. 100.
p. 122 Messerschmidt's mammoth—Breyne, pp. 137–138.
p. 122 Bones to Breyne—Vermeulen, p. 121.
p. 124 Creating the academy—Black, pp. 8–13, and Vucinich, pp. 70–72.
p. 124 Bering's first voyage—Lincoln, pp. 100–106, and Bobrick, pp. 150–156.
p. 124 Messerschmidt and Bering—Vermeulen, p. 120.
p. 124 Messerschmidt goes home—Vermeulen, p. 123.
p. 124 Strahlenberg on Messerschmidt—Strahlenberg, p. ii.
p. 124 Messerschmidt called back to St. Petersburg—Vermeulen, p. 124.
p. 125 Planning the Second Kamchatka—Bobrick, pp. 161–166, and Black, pp. 51–53.
p. 125 Academics depart—Black, pp. 53–55.
p. 125 "uncooperative officials, impassable landscapes, and Siberian mosquitoes"—Ibid., pp. 61–63.
p. 125 Friction with Bering—Ibid., p. 65.
p. 126 Müller on mammoths—Gerhard Müller, pp. 561–563.
p. 126 Gmelin's ox skulls—Gmelin, *Reise*, pp. 152–154.
p. 127 Steller background—Stejneger, pp. 8–66.
p. 127 Steller and Messerschmidt—Ibid., pp. 81–84.
p. 127 Steller goes east—Ibid., pp. 144–148.
p. 128 Failed trip to Kolyma—Ibid., pp. 81–84, and Stellero, p. 293.
p. 128 Stejneger believes—Stejneger, pp. 407–408.
p. 129 Delisle's spying—Breitfuss, pp. 90–92.

CHAPTER 6

p. 131 Flower Wars—This explanation, given by Montezuma himself, has been challenged by recent historians. See, for example, Isaac.
p. 131 Tlaxcala bone story—Díaz del Castillo, p. 286.
p. 132 Probable species—Mayor, *Fossil Legends*, p. 78.

p. 132 Mayor looks for the bone—Ibid., p. 77.
p. 132 Others in Mexico—Acosta, vol. II, pp. 453–454; Hererra y Tordesillas, fol. 23; Torrubia pp. 76–78.
p. 133 Cieza and Zárate—Cieza de Léon, pp. 189–191, Zárate, pp. 16–18.
p. 134 Witsen—Witsen, *Aeloude*, p. 3.
p. 134 *Boston News-Letter*—Stanford, p. 47.
p. 134 Cornbury's letter—Ibid., p. 48.
p. 134 Cornbury's unpopularity—Semonin, p. 17.
p. 135 Dudley to Mather—Stanford, pp. 49–50.
p. 135 Taylor on Indian reaction—Ibid., p. 53.
p. 135 Mather's low period—Levin, pp. 752–753, Semonin, pp. 27–29.
p. 136 Text of Mather's letter—Levin, pp. 761–770.
p. 137 Society's response—Mather, pp. 62–63.
p. 137 Catesby—Catesby, p. vii.
p. 138 First campaign against the Chickasaw—Atkinson, 43–61.
p. 139 Composition of Longueuil's army—"Une Expédition Canadienne," pp. 181–182.
p. 140 Shawnee reinforcements—Stevens and Kent, p. 5.
p. 141 de Lery's map—Ibid., p. 4.
p. 141 Fabry's account of the discovery—Buffon, vol. 11, p. 171
p. 141 The second campaign—Atkinson, pp. 66–73.
p. 141 Mandeville's incorrect date—Stevens, pp. 6–67.
p. 141 Bossu's incorrect date—Bossu, p. 206.
p. 142 Guettard—Guettard, pp. 349–351.
p. 142 Source of Guettard's tooth—Tassy, pp. 270–273.
p. 144 Gist and Smith—Semonin, pp. 92–94.
p. 144 Mary Ingles—Thompson, pp. 12–15.
p. 145 Shawnee peace gift—Semonin, p. 100.
p. 145 Kenny and Sutton—Hedeen, pp. 39–40.
p. 146 Collinson letters—Ibid., pp. 40–41.
p. 146 The prodigious *Mahmout*—Buffon, vol. 9, p. 126.
p. 147 Daubenton on femurs—Daubenton, pp. 207–214.
p. 147 Daubenton on teeth—Ibid., pp. 217–224.
p. 148 Elephant in *Buffon, Natural History*, vol. 11 and hippo in vol. 12.
p. 148 Turquoise teeth—Reamur.
p. 149 Croghan's trips—Semonin, pp. 104–110.
p. 149 Gordon at Big Bone Lick—Hedeen, p. 43.
p. 150 Franklin to Croghan—Semonin, p. 138.
p. 150 Franklin sends a tooth to Abbé Chappe—Ibid., p. 141.
p. 151 Collinson's letters—Collinson, "An Account," and Collinson, "Sequel."
p. 151 Hunter's researches—Hunter, pp. 34–38.
p. 152 Shelburne's questionaire—Ibid., pp. 38–40.
p. 152 Hunter's conclusions—Ibid., pp. 44–45.

CHAPTER 7

p. 153 Fossil Bones—Diederot, ed., *Encyclopédie*, vol. 7, pp. 686–687.
p. 153 Behemoth—Ibid., vol. 2, p. 191.
p. 153 Mammoth—Ibid., vol. 10, p. 7.
p. 154 Fossil Ivory—Ibid., vol. 9, pp. 63–64.
p. 155 Voltaire—Mervaud, pp. 119–120.
p. 156 Peter's orders—Cohen, pp. 64–65.
p. 157 Middendorf—Middendorff, pp. 278–279.
p. 157 Mammoth coast—Petermann map.
p. 157 Betskoi's plan and Büsching's recommendation—Black, pp. 161–162, 166.
p. 158 Pallas and Messerschmidt's journals—Vermeulen, p. 125.
p. 158 Pallas publishes Müller—Black, p. 187.
p. 158 Pallas's first paper—Pallas, "De ossibus Sibiriae."
p. 159 Müller's instructions—Black, p. 177.
p. 159 "from the Don"—Pallas, "De reliquiis," p. 576.
p. 160 Vilui rhinoceros—Pallas, *Reise durch*, pp. 98–101.
p. 160 Brandt illustration—Brandt, "Observationes."
p. 161 Camper's conclusions—Camper, "Dissertatio," pp. 202–205.
p. 161 Pallas's comments—Ibid., pp. 210–212.
p. 161 Javan rhinoceros—Rookmaaker et al., p. 125.
p. 162 Dr. Morgan's collection—Bell, p. 171.
p. 162 Michaelis views the bones—Ibid.
p. 163 Michaelis's conclusions—Ibid., pp. 173–174.
p. 163 Camper convinced—Ibid., pp. 174–175.
p. 163 Morgan sells the bones—Ibid., pp. 175–176.
p. 163 Pallas on origin of mountains—Oldroyd, pp. 83–84.
p. 164 Great eastern flood—Greene, pp. 72–73.
p. 164 Religious reaction to Buffon's geological theory—Rudwick, *Bursting*, pp. 141–142.
p. 164 The epochs—Ibid., pp. 144–147.
p. 165 Lyakhov—Cohen, p. 66.
p. 165 Buffon's experiments—Oldroyd, p. 91.
p. 166 Changes during the sixth Epoch—Buffon, Sup. 5, pp. 191–224.
p. 167 New World degeneracy—Dugatkin, chap 2.
p. 168 Marbois—Ibid., pp. 63–64.
p. 168 Jefferson on the mammoth—Jefferson, pp. 55–62.
p. 168 Jefferson and Buffon—Semonin, pp. 220–230.
p. 170 Buffon on classification schemes—Dear, pp. 50–51.
p. 170 Blumenbach categories—Blumenbach, "Einige Naturhistorische," pp. 13–24.
p. 170 Blumenbach names—Blumenbach, *Handbuch*, pp. 696–698.
p. 171 Cuvier background—Rudwick, *Georges Cuvier*, pp. 13–16.
p. 171 Geoffroy and Stadtholder's collection—Rudwick, *Bursting*, pp. 354–356.
p. 172 First version of elephant paper—Cuvier "Mémoires," in *Magasin Encyclopedique*.

p. 173 Bru and the *Megatherium*—Rudwick, *Bursting*, pp. 356–360.
p. 173 Bru and Cuvier—Rudwick, *Georges Cuvier,* p. 26, n. 1.
p. 175 Second version of elephant paper—Cuvier "Mémoires," in Journal de physique.
p. 175 Addendum—Ibid., p. 22.
p. 175 An international effort—Rudwick, Georges Cuvier, pp. 42–45.
p. 175 His own contribution—Ibid., pp. 59–63.
p. 175 The mastodon genre—"Sur le Grande Mastodonte," in Cuvier, *Recherches.*
p. 176 "bubby toothed"—Semonin, p. 355.

CHAPTER 8

p. 179 The only description of the discovery is in Adams, "Some account of a journey."
p. 180 fifty rubles in goods—In 1800, the imperial ruble was worth 1.2 grams of pure gold. As I write this the, the price of gold is down a bit; sixty grams of gold is worth about $2,400. But the difference in lifestyles between the Batouline and my neighbors is broad enough to make numbers almost meaningless. Fifty rubles was more cash than the entire community saw in five or ten years.
p. 180 Boltunov's drawing—Baer, "Fortsetzung," pp. 532–534.
p. 180 Boltunov's description—Boltunov.
p. 182 Only four more—Tolmachoff, pp. 21–23.
p. 182 Pallas's rhino—see chapter 7.
p. 182 Adams's backgtound—"Adams, Mikhail Fredrikh," *Russian Biographical Dictionary*, p. 60.
p. 182 Krusenstern on China—Krusenstern, pp. xxiv–xxxii.
p. 183 Trade at Kiakhta—Ibid., pp. xxii–xxiv.
p. 184 Golovkin's embassy departs—Anon. [Klaproth], pp. 16–23.
p. 185 Rezanov's mission—Krusenstern, pp. 281–287.
p. 185 Tilesius backgtound—De Bersaques, pp. 563–570.
p. 186 Patapov's mammoth—Tilesio, "De skeleto," 424–426.
p. 186 Golovkin held at Kiakhta—Anon. [Klaproth], pp. 26.
p. 186 Pared-down embassy—Ibid., p. 30.
p. 187 Golovkin ordered to kowtow—Ibid., pp. 41–45.
p. 187 Ordered to leave China—Timkowky and Klaproth, pp. 130–134.
p. 188 Adams in Yakutsk—Michael Adams, p. 142.
p. 189 Adams's herbarium sibiricum forwarded to the academy—Anon. "Nachtrag," p. 245.
p. 189 Shumachov takes Adams to the mammoth—Michael Adams, p. 144.
p. 189 Description of the mammoth—Ibid., p. 147.
p. 190 "[O]f such an extraordinary weight,"—Ibid., p. 148.
p. 190 Waiting for Belkoff to return—Ibid., pp. 49–50.
p. 191 Wrong tusks—Brandt, "Mittheilungen," note 2, pp. 96–97.
p. 192 Tail—Short tail, Michael Adams, p. 151. No tail, Tilesius, p. 450.

p. 193 Samples sent to colleagues—Tilesius, p. 441.
p. 193 Brain cast for Sömmerring—Ibid.
p. 194 First mention at the French academy—Cuvier, "Rapport."
p. 195 Cuvier quotes Tilesius—Cuvier, *Recherches,* 1821 ed., p. 147.
p. 196 Klaproth etymology—Tilesius, note pp. 409–411.
p. 198 JQ Adams on Tilesius—J. Q. Adams, pp. 76–77, 110, 115.
p. 199 "Adams had embellished"—Baer, "Neue Auffindung," pp. 267–268.
p. 200 Shumachov in St. Petersburg—Ibid., p. 268, n. 15.
p. 200 Maydell—quoted in Tolmachoff, p. 16.
p. 200 Foca—Digby, pp. 82–84.
p. 201 Bunge—Tolmachoff, p. 31.
p. 202 DNA—Gilbert, Thomas, et al.

AFTERWORD

p. 204 Gmelin—*Flora*, pp. LXV–LXX.
p. 204 Shergin's well—Barry, p. 165.
p. 204 Most studied—for example, Siegelt et al.
p. 204 Kerr—Rudwick, *Bursting,* p. 510.
p. 205 Falconer's classification—Cohen, pp. 136–137.
p. 205 Later classifications—Tassy and Shoshani.
p. 205 Lartet—Cohen, pp. 152–154.
p. 206 Blainville—Blainville, "Mémoire sur les ossements" and "Sur les ossements."
p. 206 Vrba—Scigliano, pp. 32–33.

BIBLIOGRAPHY

BOOKS AND ARTICLES

A.M. [Allen Mullen] *An Anatomical Account of the Elephant Accidentally Burnt in Dublin, on Friday, June 17, in the Year 1681.* London: Sam. Smith, 1681.

Abel, Othenio. *Die Tiere der Vorwelt.* Leipzig and Berlin: B. G. Teubner, 1914.

Abel, Othenio. *Die Vorweltlichen Tiere in Märchen, Asge und Aberglauben.* Karlsruhe: G. Braunsche Hofbuchdruckerei und Verlag, 1923.

Abel, Othenio. *Lebensbilder aus der tierwelt der vorzeit.* Jena: Vearlag von Gustav Fischer, 1922.

Acosta, José de. *The Natural & Moral History of the Indies*, trans. and ed. Clements R. Markham. London: Hakluyt Society, 1880.

Adams, John Quincy. *Memoirs of John Quincy Adams: Comprising Portions of His Diary from 1795 to 1848*, vol. 2, ed. Charles Francis Adams. Philadelphia: J. B. Lippincott, 1874.

Adams, Michael. "Some account of a journey to the frozen sea, and of the discovery of the remains of a mammoth. Translated from the French." *Philosophical Magazine*, ser. 1, vol. 29, no. 114 (Nov. 1807): 141–153.

"Adams, Mikhail Fredrikh." *Russian Biographical Dictionary*, vol. 1. St. Petersburg: I. N. Skorokhodova, 1896.

Agnesi, V., C. Di Patti, and B. Truden. "Giants and elephants of Sicily." *Geological Society, London, Special Publications* 273 (2007): 263–270.

Ahlenius, Karl Jakob Mauritz. *Olaus Magnus och hans framställning af Nordens geografi, studier i geografiens historia.* Upsala: Almquist & Wikselles Boktryckery-Arktibolag, 1895.

Angerer, Leonard P. "Die Wiederauffindung der von den Schweden im Jahre 1645 zu Krems in Nieder-österreich aus gegrabenen Mammut knochen in der Stifts-sammlung von Kremsmünster." *Verhandlungen der k. k. geologischen Reichsanstalt* 16 (1911): 359–360.

Anon. [Julius Klaproth]. *Die russische Gesandtschaft nach China im Jahr 1805.* St. Petersburg: Ziemsenschen Verlag, 1809.

Anon. *The King's Mirror (Speculum regale-Konungs skuggsjá) translated from the Old Norwegian by Laurence Marcellus Larson.* New York: New York American-Scandinavian Foundation, 1917.

Anon. [Johann Christoph Schnetter]. *Kurtze doch ausführliche Beschreibung Des Unicornu Fossilis, oder gegrabenen Einhorns, Welches in der Herschafft Tonna gefunden worden, Verfertiget von dem Colligio Medico in Gotha, den 14. Febr. 1696.* Gotha: Christoph Reyhern, 1696.

Anon. "Nachtrag zu den authentischen Nachrichten von der Russischen Gesandtschaft nach China, in den Jahren 1805 u. 1806." *Allgemeine geographische Ephemeriden* 25 (Feb. 1808): 243–247.

Anon. *The Present State of Europe: The Historical and Political Monthly Mercury* 7, no. 12 (Dec. 1696): 402–403.

Anon. Review of Wilhelmi Ernesti Tenzilii, "Epistola de Sceleto Elephantino Tonnæ nuper Efosso, ad Virum toto orbe Celeberrimum Antonium Magliabechium Serenissimi Magni Ducis Hetruriæ Bibliothecarium & Consiliarium." *Journal des sçavans* (Amsterdam ed.) 54 (1696): 595–599.

Ariew, Roger. "Leibniz on the Unicorn and Various Other Curiosities." *Early Science and Medicine*, 3–4 (1998): 267–288.

Armstrong, T. "In Search of a Sea Route to Siberia, 1553–1619." *Arctic* 37, no. 4 (Dec. 1984): 429–440.

Armstrong, Terence. *Russian Settlement in the North.* Cambridge: Cambridge University Press, 1965.

Atkinson, James R. *Splendid Land, Splendid People: The Chickasaw Indians to Removal.* Tuscaloosa: University of Alabama Press, 2004.

Avril, Philippe. *Voyage en divers etats d'Europe et d'Asie, entrepris pour découvrir un nouveau chemin à la Chine.* Paris: Chez Jean Boudot, 1693.

Baer, K. E. von. "Anatomische und zoologische Untersuchungen über das Wallross (Trichechus rosmarus) und Vergleichung dieses Thiers mit andern See-Säugethieren." *Mémoires de l'Académie impériale des sciences de St. Pétersbourg. Sér. 6. Sciences mathématiques, physiques et naturelles* 2 (1838): 97–235.

Baer, K. E. von. "Fortsetzung der Berichte über die Expedition zur Aufsuchung des angekündigten Mammuths." *Bulletin De L'academie Imperiale Des Sciences* 10 (1866): 513–534.

Baer, K. E. von. "Neue Auffindung eines vollständigen Mammuths, mit der Haut und den Weichtheilen, im Eisboden Sibiriens, in der Nähe der Bucht des Tas." *Bulletin De L'academie Imperiale Des Sciences* 9 (1866): 230–296.

Bagrow, Leo. "Semyon Remezov, A Siberian Cartographer." *Imago Mundi* 11 (1954): 111–125.

Bagrow, Leo. "Sparwenfeld's Map of Siberia." *Imago Mundi* 4 (1947): 65–70.

Balthasar, Joseph Anton Felix von. *Nachrichten von der Stadt Luzern und ihrer Regierungsverfassung, oder historische und moralische Erklärungen der acht ersten Gemälde auf der Kapellbrücke der Stadt Luzern.* Luzern: Salzmann, 1784.

Barry, Roger, and Thian Yew Gan. *The Global Cryosphere: Past, Present, and Future.* Cambridge: Cambridge University Press, 2011.

Bartholin, Thomas. *De unicornu observationes novæ.* Amsterdam: J. Henr. Wetstenium, 1678.

Beer, Rüdiger Robert. *Unicorn: Myth and Reality.* New York: Mason/Charter, 1977.

Behre, Karl-Ernst. "Coastal Development, Sea-Level Change and Settlement History during the Later Holocene in the Clay District of Lower Saxony (Niedersachsen), Northern Germany." *Quaternary International* 112 (2004): 37–53.

Behrens, Georg Henning. *Hercynia Curiosa oder Curiöser Hartz-Wald: Das ist Sonderbahre Beschreibung und Verzeichnis Derer Curiösen Hölen, Seen, Brunnen, Bergen, und vielen andern an-und auff dem Hartz . . .* Nordhausen, 1703.

Bell, John. *Travels from St. Petersburg in Russia to Various Parts of Asia*, 2 vols. Edinburgh: Robert and Andrew Foulis, 1763.

Bell, Whitfield J., Jr. "A Box of Old Bones: A Note on the Identification of the Mastodon." *Proceedings, American Philosophical Society* 93, no. 2 (May 16, 1949): 169–178.

Bilfinger, G. B. *Varia in fasciculos collecta*. Stuttgardiae: Sumtibus filiorum Beati Christophori Erhardti, 1743.

Bjørnbo, Axel Anton. "Cartographica Groenlania." *Meddelelser om Grønland* 48 (1912): 1–332.

Black, Joseph Lawrence. *G. F. Müller and the Imperial Russian Academy of Sciences, 1725–1783*. Kingston-Montréal: McGill-Queen's University Press, 1986.

Blainville, Henri Ducrotay de. "Mémoire sur les ossements fossiles attribués au prétendu géant Theutobochus, roi des Cimbres." *Nouvelles Annales du Muséum national d'histoire naturelle* 4 (1835): 37–74.

Blainville, Henri Ducrotay de. "Sur les ossements fossiles attribués au prétendu géant Teutobochus." *Comptes rendus hebdomadaires des séances de l'Académie des sciences* 4 (1837): 633–634.

Blumenbach, Johann. "Einige Naturhistorische Bemerkungen bey Gelegenheit einer Schweizer-Reise." *Magazin für das Neueste aus der Physik und Naturgeschichte* 5 (1788): 13–22.

Blumenbach, Johann. *Handbuch der Naturgeschichte*. Göttingen: Johann Christian Dieterich, 1799.

Bobrick, Benson. *East of the Sun: The Epic Conquest and Tragic History of Siberia*. New York: Poseidon Press, 1992.

Boltunov, Roman. "Description of the animal called the Mammoth found in Izhiganskoy county, Yakutsk." *Technical Journal* [in Russian] 3, no. 4 (1806): 162–166.

Borgnet, Auguste, ed. *B. Alberti Magni Ratisbonensis episcopi, ordinis Prædicatorum, Opera Omnia: Ex editione lugdunensi religiose castigata*, vol. 12. Paris: Ludovicum Vives, 1891.

Bossu, Jean-Bernard. *Nouveaux voyages aux Indes occidentales*, vol. 1. Paris: Chez Le Jay, 1763.

Bosworth, Joseph, ed. *A Description of Europe, and the Voyages of Othere and Wulfstan, Written in Anglo-Saxon by King Alfred the Great*. London: Longman and Co., 1855.

Bourguet, Louis. *Traité des pétrifications*. Paris: Chez Briasson, 1742.

Brand, Adam. *Beschreibung Der Chinesischen Reise: Welche vermittelst Einer Zaaris Besandschaft Durch Dero Ambassadeur Hern Isbrand*. Hamburg: Benjamin Schillern, 1698.

Brandt, J. F. "Kurze Bemerkungen Über Aufrechtstehende Mammuthleichen." *Bulletin de la Société impériale des naturalistes de Moscou* 40, no. 3 (1867): 241–256.

Brandt, J. F. "Mittheilungen über die Gestalt und Unterscheidungsmerkmale des Mammuth oder Mamont (Elephas primigenius)." *Bulletin de L'academie Imperiale des Sciences* 10 (1866): 93–118.

Brandt, J. F. "Observationes ad rhinocerotis tichorhini historiam spectantes." *Mémoires de l'Académie impériale des sciences de St. Pétersbourg, Sér. 6, Sciences naturelles* 7 (1849): 161–289.

Breitfuss, L. "Early Maps of North-Eastern Asia and of the Lands around the North Pacific. Controversy between G. F. Müller and N. Delisle." *Imago Mundi* 3 (1939): 87–99.

Breyne, Johann Philip. "A Letter from John Phil. Breyne, M. D. F. R. S. to Sir Hans Sloane, Bart. Pres. R. S. with Observations and a Description of some Mammoth's Bones dug up in Siberia, proving them to have belonged to Elephants." *Philosophical Transactions of the Royal Society* 40 (1737–1738): 124–139.

Brückmann, Franz Ernst. *Epistola itineraria XII: De Gigantum dentibus ad virum nobilissimum, atque doctissimum dominum . . .* Wolfenbüttel: 1729.

Bruman, Henry J. "The Schaffhausen Carta Marina of 1531." *Imago Mundi* 41 (1989): 124–132.

Buffetaut, Eric. *Short History of Vertebrate Paleontology*. New York: Springer Verlag, 1987.

Buffon, Georges Louis Leclerc. *Histoire naturelle, générale et particulière: Avec la description du Cabinet du Roi*. 36 vols. Paris: 1748–1788.

Cahen, Gaston. *Les cartes de la Sibérie au XVIIIe siècle, essai de bibliographie critique*. Paris: Imprimerie nationale, 1911.

Camper, Peter. "Complementa varia Acad. Imperiali Scient Petropolitanae communicanda ad Clar, ac Celeb. Virum P. S. Pallas." *Nova acta Academiae scientiarum imperialis petropolitanae* 3 (1784): 250–267.

Camper, Peter. "Description Anatomique d'un Elephant mâle." In *Oeuvres de Pierre Camper*, vol. 2, pp. 21–282. Paris: H. J. Janson, 1803.

Camper, Petro. "Dissertatio de Cranio Rhinocerotis Africani, Cornu Gemino; Academiae Scientiarum Imperiali Petropolitanae Oblata." *Acta Academiae scientiarum imperialis petropolitanae*, part 2 (1777): 193–221.

Catesby, Mark. *The natural history of Carolina, Florida, and the Bahama Islands: Containing the Figures of birds, Beasts, Fishes, Serpents, Insects, and Plants*, vol. 2. London: Mark Catesby, 1743.

Céard, Jean. "La querelle des géants et la jeunesse du monde." *Journal of Medieval and Renaissance Studies* 8, no. 1 (1978): 37–76.

Christian, David. "Silk Roads or Steppe Roads? The Silk Roads in World History." *Journal of World History* 11, no. 1 (2000): 1–26.

Cieza de Léon, Pedro. *The Travels of Pedro de Cieza de Léon, A.D. 1532–50*, trans. and ed. Clements R. Markham. London: Hakluyt Society, 1874.

Cohen, Claudine. *The Fate of the Mammoth: Fossils, Myth, and History*. Chicago: University of Chicago Press, 2003.

Collet, Dominik. *Die Welt in der Stube: Begegnungen mit Außereuropa in Kunstkammern der Frühen Neuzeit*. Göttigen: Vandenhoeck & Ruprecht GmbH & Co., 2007.

Collinson, Peter. "An Account of some very large Fossil Teeth found in North America, and described by Peter Collinson, F.R.S." *Philosophical Transactions of the Royal Society* 57 (1767): 464–467.

Collinson, Peter. "Sequel to the foregoing Account of the large Fossil Teeth." *Philosophical Transactions of the Royal Society* 57 (1767): 468–478.

Cumston, Charles Greene. "Jacques Tissot's Brochure Entitled 'Discours Veritable de la Vie, Mort et Des Os du Geant Theutobocus.' A Contribution to the History of Osteology." *Post-Graduate* 27, no. 8 (August 1912): 665–679.

Cutler, Alan. *The Seashell on the Mountaintop: A Story of Science, Sainthood, and the Humble Genius Who Discovered a New History of the Earth*. New York: Dutton, 2003.

Cuvier, Georges. "Mémoires sur les espèces d'éléphants vivants et fossiles." *Journal de physique, de chimie, d'histoire naturelle et des arts* 50 (1800): 207–217.

Cuvier, Georges. "Mémoires sur les espèces d'éléphants vivants et fossiles." *Magasin Encyclopedique* 2, no. 3 (1796): 440–445.

Cuvier, Georges. "Rapport à la Classe des Sciences Physiques et Mathématiques de l'Institut." *Annales du Muséum d'histoire naturelle* 10 (1807): 381–386.

Cuvier, Georges. *Recherches sur les ossemens fossiles de quadrupèdes, où l'on rétablit les caractères de plusieurs espèces d'animaux que les révolutions du globe paroissent avoir détruites*. Paris: Chez Deterville, 1812.

Cysat, Johann Leopold. *Beschreibung dess berühmbten Lucerner-oder 4. Waldstaetten Sees und dessen fürtrefflichen Qualiteten und sonderbaaren Eygenschafften*. Luzern: David Hautten, 1661.

Darwin, Charles. *Narrative of the Surveying Voyages of His Majesty's Ships Adventure and Beagle: Between the Years 1826 and 1836*, vol. 3. London: Henry Colburn, 1839.

Daubenton, Louis-Jean-Marie. "Memoire sur des os et des dents remarquables par leur grandeur." *Histoire de l'Académie Royale des Sciences, Memoires de Mathématique et de Physique* 1762 (1764): 206–230.

Dawkins, Boyd. "On the Range of the Mammoth." *Popular Science Review* 7 (1868): 275–286.

Dear, Peter. *The Intelligibility of Nature: How Science Makes Sense of the World.* Chicago: University of Chicago Press, 2006.

De Bersaques, J. "Wilhelm Gottlieb Tilesius—a Forgotten Dermatologist." *Journal der Deutschen Dermatologischen Gesellschaft* 9 (2011): 563–570.

De Bruyn, Cornelius. "An Abstract of M. Cornelius Le Brun's Travels Through Russia and Persia to the East Indies Containing the Observations He made in Russia." In *The Present State of Russia*, ed. Friedrich Christian Weber, vol. 2. London: W. Taylor, 1723.

Desfontaines, Pierre François Guyot. "Voici quelques Mémoires que j'ai recus de Dauphiné, au sujet du Géant Theutobochus, dont ilest parlé," Tome III. *Jugemens sur quelques ouvrages nouveaux* 6 (1744): 217–320.

Díaz del Castillo, Bernal. *The True History of the Conquest of New Spain*, vol. 1, trans. and ed. Alfred Percival Maudslay. London: Hakluyt Society, 1908.

Diderot, Denis, ed. *Encyclopédie: Ou dictionnaire raisonné des sciences, des arts et des métiers*, 28 vols. Paris and Neufchatel: Samuel Faulche, 1751–1772.

Digby, Basset. *The Mammoth and Mammoth-Hunting in Northeast Siberia.* New York: D. Appleton, 1926.

Dugatkin, Lee Alan. *Mr. Jefferson and the Giant Moose: Natural History in Early America.* Chicago: University of Chicago Press, 2009.

Dunér, David. "On the Decimal: The First Russian Translation of Swedenborg." *New Philosophy* (July–December 2009): 933–944.

Ehrensvärd, Ulla. "Die Sibirienkarte des Philipp Johann von Strahlenberg (1730) und ihre Bedeutung für das moderne Kartenbild vom nördlichen Asien." *Cartographica Helvetica: Fachzeitschrift für Kartengeschichte* 43–44 (2011): 17–33.

Eiseley, Loren. *The Firmament of Time.* New York: Bison Books, 1999 [1960].

Falconer, Hugh. *Fauna Antiqua Sivalensis, Being the Fossil Zoology of the Sewalik hills in the North of India, Part 1, Proboscidea.* London: Smith, Elder and Co., 1846.

Finch, Richard. "A Letter of Richard Finch to the Right Worshipfull Sir Thomas Smith, Governour; and to the rest of the Worshipfull Companie of English Merchants, trading into Russia: touching the former Voyag, and other observations." In *Hakluytus posthumus, or, Purchas his Pilgrimes: contayning a history of the world in sea voyages and lande travells by Englishmen and others*, vol. 13, p. 214. 1906.

Findlen, Paula. *Possessing Nature: Museums, Collecting, and Scientific Culture in Early Modern Italy.* Berkeley: University of California Press, 1994.

Fisher, Raymond H. *The Russian Fur Trade, 1550–1700.* Berkeley: University of California Press, 1943.

Florschütz, G. "Die erste Aufdeckung des Elephas antiquus in den in Sandbrüchen bei Gräfentonna." *Mitteilungen der Vereinigung für Gothaische Geschichte und Altertumsforschung* (1905): 43–57.

Forssell, H. L. "Minne af erkebiskopen Erik Benzelius den yngre." *Svenska akademiens handlingar*, part 58, pp. 113–476. Stockholm, 1883.

Fournier, Édouard. "Notes on 'Discours veritable de la vie, mort, et des os du Geant Theutobocus, roy des Theutons, Cimbres et Ambrosins, lequel fut deffaict 105 ans avant la venue de nostre Seigneur Jesus-Christ,' by Jacques Tissot de Tournon." *Variétés historiques et littéraires* 9 (1859): 241–259.

Fraas, Oscar. *Vor der Südfluth! Eine Geschichte der Urwelt.* Stuttgart: Hoffmann'sche Verlags-Buchhandlung, 1866.

Galbreath, C. B., ed. *Expedition of Celoron to the Ohio Country in 1749.* Columbus, OH: F. J. Heer Printing Co., 1920.

Gambaccini, Piero. *Mountebanks and Medicasters: A History of Italian Charlatans from the Middle Ages to the Present.* Jefferson, NC: McFarland & Company, 2004.

Gandilhon, René. "L'éléphant de Claude de Lorraine (1628)." *Bibliothèque de l'école des chartes* 114 (1956): 208–211.

Gassendi, Pierre. *The Mirrour of True Nobility & Gentility Being the Life of Peiresc, Englished by W. Rand.* London: J. Streater, 1657.

Gaudant, Jean. "Histoire d'une brève controverse: Wilhelm Ernst Tentzel (1659–1707) et l'éléphant fossile découvert en 1695 à Burg Tonna, près de Gotha (Allemagne)." *Travaux du Comité Français d'Histoire de la Géologie (COFRHIGÉO)*, Troisième série, 24, no. 6 (Dec. 8, 2010): 117–130.

Gayarré, Charles. *Louisiana: Its Colonial History and Romance*, vol. 2. New York: Harper & Brothers, 1851.

Gebhard, Johan Fredrik, ed. *Het leven van Mr. Nicolaas Cornelisz. Witsen*, vol. 2. Utrecht: J. W. Leeflang, 1882.

George, Wilma. *Animals and Maps.* Berkeley: University of California Press, 1969.

Gesnerum, Conradum [Konrad Gesner]. *Nomenclator aquatilium animantium. Icones animalium aquatilium in maris & dulcibus aquis degentium, plusquam DCC. cum nomenclaturis singulorum Latinis, Greçis, Italicis, Hispanicis, Gallicis, Germanicis, Anglicis, alji'sq; interdum per certos ordines digestae.* Tiguri: Excudebat Christoph. Froschoverus, 1560.

Gilbert, M., P. Thomas, et al. "Intraspecific phylogenetic analysis of Siberian woolly mammoths using complete mitochondrial genomes, Supporting Information." *Proceedings of the National Academy of Sciences of the USA* 105, no. 24 (2008).

Gillio, Petro. *Descriptio Nova Elephanti.* Hamburg: Bibliopolio Heringiano, 1614.

Ginsburg, Leonard. "Nouvelles lumières sur les ossements fossiles autrefois attribués au géant Theutobochus." *Annales de Paléontologie* [Paris] 70 (1984): 181–219.

Gmelin, Johann Georg. *Flora sibirica, sive Historia plantarum Sibiriae*, vol. 1. St. Petersburg: Academy of the Sciences, 1747.

Gmelin, Johann Georg. *Reise durch Sibirien, von dem Jahr 1738 bis zu Ende 1740*, vol. 3. Göttengen: Vandenhoeck, 1752.

Godard, Gaston. "The fossil proboscideans of Utica (Tunisia), a key to the 'giant' controversy, from Saint Augustine (424) to Peiresc (1632)." *Geological Society, London, Special Publications* 310 (2009): 67–76.

Gould, Stephen Jay. *Time's Arrow, Time's Cycle: Myth and Metaphor in the Discovery of Geological Time.* Cambridge, MA: Harvard University Press, 1988.

Greene, John C. *The Death of Adam: Evolution and Its Impact on Western Thought.* Ames: Iowa State University Press, 1996.

Guericke, Otto von. *Experimenta Nova (ut vocantur) Magdeburgica de Vacuo Spatio.* Amstelodami: Joannem Janssonium à Waesberge, 1672.

Guettard, Jean Etienne. "Memoire dans lelquel on compare le Canada à la Suisse par rapport à ses minéraux, seconde partie." *Histoire de l'Académie Royale des Sciences, Memoires de Mathématique et de Physique* 1752 (1756): 323–360.

Habicot, Nicholas. *Antigigantologie Contre Discours De La Grandeur Des Geans.* Paris: Chez Jean Corrozet, 1618.

Hag, Torgny. "Karoliner och Behemoter: 1700-talets svenska diskussion om mammuten." *Svenska Linnésällskapets årsskrift* (1982): 51–72.

Happelius [Eberhard Werner Happel]. *Größte Denkwürdigkeiten der Welt oder so genannte Relationes curiosae*, vol. 4. Hamburg: Thomas von Wiering, 1689.

Hase, Theodor. *Dissertationis Septimae de Manmuth, sive Maman, Quod animal in regionibus septentrionalibus sub terram vivere referunt, in Dissertationum et observationum philologicarum sylloge.* Bremen: Sumt. Hermanni Jaegeri, 1731.

Heaney, Michael. "The Implications of Richard James's maimanto." *Oxford Slavonic Papers* 9 (1976): 102–109.

Hedeen, Stanley. *Big Bone Lick: The Cradle of American Paleontology.* Lexington: University Press of Kentucky, 2008.

Helimski, Eugen. "Etymological Notes." In *Research on Historical Grammar and Lexicology* (in Russian), pp. 30–42. Moscow, 1990.

Herberstein, Sigmund. *Notes upon Russia: Being a translation of the earliest account of that country, entitled Rerum moscoviticarum commentarii*, trans. and ed. R. H. Major, 2 vols. London: Hakluyt Society, 1851.

Herrera y Tordesillas, Antonio de. *Novus Orbis, sive, Descriptio Indiae Occidentalis.* Amstelodami: Michaelem Colinium, 1622.

Hofstra, Tette, and Kees Samploius. "Viking Expansion Northwards: Mediaeval Sources." *Arctic* 48, no. 3 (Sept. 1995): 235–247.

Hooke, Robert. "A Discourse on Earthquakes." In *The Posthumous Works of Robert Hooke, M.D. S.R.S. Geom. Prof. Gresh. &c. Containing his Cutlerian Lectures*, pp. 279–450. London: Sam. Smith and Benj. Waterford, 1705.

Howarth, H. H. *The Mammoth and the Flood.* London: Gilbert and Rivington, 1887.

Hughes, Lindsey. *Sophia: Regent of all the Russias, 1657–1704.* New Haven, CT: Yale University Press, 1990.

Hundt, Michael. *Beschreibung der dreijährigen chinesischen Reise: Die russische Gesandtschaft von Moskau nach Peking 1692 bis 1695 in den Darstellungen von Eberhard Isbrand Ides und Adam Brand.* Stuttgart: Franz Steiner Verlag, 1999.

Hunter, William. "Observations on the Bones, Commonly Supposed to Be Elephant Bones, Which Have Been Found Near the River Ohio in America." *Philosophical Transactions of the Royal Society* 58 (1768): 34–45.

Ides, Evert Ysbrants. *Three years travels from Moscow over-land to China: thro' Great Ustiga, Siriania, Permia, Sibiria, Daour, Great Tartary, etc. to Peking.* London: W. Freeman, 1705.

Isaac, Barry L. "The Aztec 'Flowery War': A Geopolitical Explanation." *Journal of Anthropological Research* 39, no. 4 (1983): 415–432.

Ivanov, V. N. "Tatshchev's Mammoth." *From the History of Biology* (in Russian) 4 (1973): 209–217.

Jefferson, Thomas. *Notes on the State of Virginia.* London, 1784.

Jones, Steve. *Revolutionary Science: Transformation and Turmoil in the Age of the Guillotine.* New York: Pegasus Books, 2017.

Keuning, Johannes. "Nicolaas Witsen as a Cartographer." *Imago Mundi* 11 (1954): 95–110.

Kiparsky, V. "Das Mammut." *Zeitschrift für Slavische Philologie* 26 (1958): 296–301.

Kircher, Athanasius. *Mundus Subterraneus*, vol. 2. Amsterdam: Joannem Janssonium à Waesberge & filios, 1678.

Knoespel, Kenneth J. "The Edge of Empire: Rudbeck and Lomonosov and the Historiography of the North." *In Search of an Order: Mutual Representations in Sweden and Russia during the Early Age of Reason*, ed. Ulla Birgergård and Irina Sandomirskaja. Stockholm: Södertörn University, 2004.

Krafft, Fritz. "Gottfried Wilhelm Leibniz oder Otto von Guericke?—Protogaea oder Experimenta nova Magdeburgica? Die Rekonstruktion des vermeintlichen Einhorns von Quedlinburg." *Sudhoffs Archiv* 99 (2015): 166–208.

Krusenstern, Adam Johann von. *Voyage round the world, in the years 1803, 1804, 1805, & 1806.* London: John Murray, 1813.

Lambecius, Petrus [Peter Lambeck]. *Commentariorum de augustissima Bibliotheca Caesarese Vindobonensi*, vol. 6. Vienna: Iohannis Thomae Trattner, 1674.

Lange, Lorenz. "Journal of Laurence Lange's Travels to China." In *The Present State of Russia*, vol. 2, ed. Friedrich Christian Weber. London: W. Taylor, 1723.

Laufer, Berthold. "Arabic and Chinese Trade in Walrus and Narwhal Ivory." *T'oung pao* 14 (1914): 315–364.

Laufer, Berthold. *Ivory in China.* Anthropology Leaflet 21. Chicago: Field Museum of Natural History, 1925.

Laufer, Berthold. "Supplementary notes on walrus and narwhal ivory." *T'oung pao* 17 (1916): 348–389.

Lavers, C. and M. Knapp. "On the Origin of Khutu." *Archives of Natural History* 35 (2008): 306–318.

Levin, David. "Giants in the Earth: Science and the Occult in Cotton Mather's Letters to the Royal Society." *William and Mary Quarterly*, 3rd ser., 45, no. 4 (1988): 751–770.

Leibniz, Gottfried Wilhelm. *Protagea*, trans. Claudine Cohn and Andre Wakefield. Chicago: University of Chicago Press, 2008.

Leibniz, Gottfried Wilhelm. *Sämtliche schriften und briefe, Series I, Allgemeiner, politischer und historischer Briefwechsel*, vols. 12 and 13. Berlin: Akademie Verlag, 1987–1990.

Lincoln, Bruce. *The Conquest of a Continent: Siberia and the Russians.* Cornell, NY: Cornell University Press, 2007.

Lister, Adrian, and Paul Bahn. *Mammoths.* New York: Macmillan, 1994.

Logan, Josias. "Extracts taken out of two Letters of Josias Logan from Pechora to Master Hakluyt Prebend of Westminster." In *Hakluytus posthumus, or, Purchas his Pilgrimes: Contayning a history of the world in sea voyages and lande travells by Englishmen and others*, vol. 13, p. 236. 1906.

Love, Ronald S. "In Search of a Passage to China: Philippe Avril's Quest for Grand Tartary." In *Distant Lands and Diverse Cultures: The French Experience in Asia, 1600–1700.* Westport, CT: Praeger, 2003.

Lovejoy, Arthur O. *The Great Chain of Being: A Study in the History of an Idea.* Cambridge, MA: Harvard University Press, 1936.

Ludolfi, Henrici Wilhelmi [Heinrich Wilhelm Ludolf]. *Grammatica Russica.* Oxford, 1696.

Luffkin, John. "Part of a Letter from Mr John Luffkin to the Publisher, concerning Some Large Bones, Lately Found in a Gravel-Pit Near Colchester." *Philosophical Transactions of the Royal Society* 22 (1700): 924–926.

Lydekker, R. "Mammoth Ivory." In *Annual Report of the Board of Regents of the Smithsonian Institution.* Washington, DC: Government Printing Office, 1901.

Magnus, Olaus. *Historia de gentibus septentrionalibus, earumque diversis statibus, conditionibus, moribus, ritibus . . . nec non universis pene animalibus in Septentrione degentibus, eorumque natura.* Rome: Giovanni M. Viotto, 1555.

Massie, Robert K. *Peter the Great: His Life and World.* New York: Modern Library, 2012.

Mather, Cotton. "An Extract of Several Letters from Cotton Mather, D. D. to John Woodward, M.D. and Richard Waller, Esq; S. R. Secr." *Philosophical Transactions of the Royal Society* 29 (1714): 62–71.

Mayor, Adrienne. *The First Fossil Hunters: Paleontology in Greek and Roman Times.* Princeton, NJ: Princeton University Press, 2001.

Mayor, Adrienne. *Fossil Legends of the First Americans.* Princeton, NJ: Princeton University Press, 2007.

McTavish, Lianne. *Childbirth and the Display of Authority in Early Modern France.* Burlington, VT: Ashgate, 2005.

Meredith, Martin. *Elephant Destiny: Biography of an Endangered Species in Africa.* New York: Public Affairs, 2001.

Merian, Matthew. *Theatrum Europaeum, oder außführliche und warhafftige Beschreibung aller und jeder denckwürdiger, vom A. 1642 bis A. 1647,* vol. 5. Franckfurt: Wolffgang Hoffmans, 1651.

Mervaud, Michel. "Un monstre siberien dans l'Encyclopedie, et ailleurs: Le Behemoth." *Recherches sur Diderot et sur l'Encyclopedie* 17 (October 1994): 107–132.

Messerschmidt, Daniel Gottlieb. "Nachricht von D. Daniel Gottlieb Messerschmidts Siebenjahriger Reise in Siberien, Peter Simon Pallas, ed." *Neue nordische Beyträge zur physikalischen und geographischen Erd-und Völkerbeschreibung, Naturgeschichte und Oekonomiee* 3 (1782): 97–159.

Meulen, Reinder van der. *De Naam van den mammouth* (Mededeelingen der Koninklijke Akademie van Wetenschappen. Afd. Letterkunde. dl. 63. ser. A. no. 12.): 349–403.

Michow, Heinrich. *Die ältesten Karten von Russland: Ein beitrag zur historischen Geographie.* Hamburg: L. Friedrichsen & Co., 1884.

Middendorff, Alexander Theodor von. *Sibirische Reise: Uebersicht der Natur Nord-und Ost-Sibiriens, Part 4, Vol. 1, Orographie und Geognosie.* Eggers, 1860.

Miechowa, Maciej. *Tractatus de duabus Sarmatis Europiana et Asiana et de contentis in eis.* Vienna: Grimm u. Wirsung, 1518.

Molyneux, Thomas. "A Discourse Concerning the Large Horns Frequently Found under Ground in Ireland, Concluding from Them That the Great American Deer, Call'd a Moose, Was Formerly Common in That Island: With Remarks on Some Other Things Natural to That Country." *Philosophical Transactions of the Royal Society* 19 (1695): 489–512.

Molyneux, Thomas. "Part of 2 Letters from Mr. Thomas Molyneux concerning a Prodigious Os Frontis in the Medicine School at Leyden. Dec. 29th. 1684. and Febr. 13th. 1685." *Philosophical Transactions of the Royal Society* 15 (1685): 880–881.

Molyneux, Thomas. "Remarks Upon the Aforesaid Letter and Teeth, by Thomas Molyneux, M. D. and R. S. S. Physician to the State in Ireland: Address'd to His Grace the Lord Archbishop of Dublin." *Philosophical Transactions of the Royal Society* 29 (1714): 370–384.

Mommsen, Theodor. *The History of Rome,* vol. 3, trans. Rev. William P. Dickson. London: Eichard Bently, 1863.

Morand, Sauveur François. "Sur la Vie, & les Ecrits de Habicot." In *Opuscules de chirurgie,* pp. 99–113. Paris: Chez Guillaume Desprez, 1768.

Mowat, Farley. *Sea of Slaughter.* Boston: Atlantic Monthly Press, 1984.

Müller, Gerhard Friedrich von. *Sammlung Russischer Geschichte* 3, no. 5–6 (1760): 561–563.

Müller, Gottfried Adrian. "Beschreibung und Abbildung einiger in dem Kabinett des Herrn geheimen Finanzraths, Gottfried Adrian Müller, befindlichen und ehedem bey Quedlinburg ausgegrabenen Knochen eines ausländischen Thieres." *Beschäftigungen der Berlinischen Gesellschaft naturforschender Freunde* 2 (1776): 340–347.

Müller, Johann Bernard. "The Manner and Customs of the Ostiaks." In *The Present State of Russia*, vol. 2, ed. Friedrich Christian Weber. London: W. Taylor, 1723.

Nevile, Francis. "A Letter of Mr. Francis Nevile to the Right Reverend St. George Lord Bishop of Clogher, R. S. S. Giving an Account of Some Large Teeth Lately Dugg up in the North of Ireland, and by His Lordship Communicated to the Royal-Society." *Philosophical Transactions of the Royal Society* 29 (1714): 367–370.

Nigg, Simon. *Sea Monsters: A Voyage around the World's Most Beguiling Map*. Chicago: University of Chicago Press, 2013.

Nordenskiold, A. E. "An Old Drawing of a Mammoth." *Nature* 819, no. 32 (July 9, 1885): 228–229.

Okabe, Shoichi. "Russian Grammars before Lomonosov." *Kanazawa University Department of Literature Essays, Literature Review* (1985): 117–142.

Oldroyd, David. *Thinking about the Earth: A History of Ideas in Geology*. London: Athlone Press, 2000.

Pallas, P. S. "De reliquiis animalium exoticorum per Asiam borealem repertis complementum." *Novi commentarii Academiae scientiarum imperialis Petropolitanae* 17 (1772): 576–606.

Pallas, P. S. "Observatio de Dentibus Molaribus Fossilibus Ignoti Animalis, Canadensibus Analogis, Etiam ad Uralense Jugum Repertis." *Acta Academiae scientiarum imperialis petropolitanae*, part 2 (1777): 213–222.

Pallas, Peter Simon. "De ossibus Sibiriae fossilibus craniis praesertim rhinocerotum atque buffalorum, observationes." *Novi commentarii Academiae scientiarum imperialis Petropolitanae* 13 (1769): 436–477.

Pallas, Peter Simon. *Reise durch verschiedene Provinzen des russischen Reichs in einem ausfuehrlichen Auszuge*, vol 3. Frankfurt and Leipzig: J. G. Fleischer, 1778.

Pasquier, Estienne. *Les Recherches de la France*. Paris: Pierre Menard, 1643.

Peale, Rembrandt. *An Historical Disquisition on the Mammoth: or, Great American Incognitum, an Extinct, Immense, Carnivorous Animal, Whose Fossil Remains Have Been Found in North America*. London: C. Mercier, 1803.

Pennant, Thomas. *Synopsis of Quadrupeds*. Chester: J. Monk, 1771.

Perry, John. *The State of Russia under the Present Czar*. London: B. Tooke, 1716.

Peters, Marion H. *De Wijze Koopman: Het wereldwijde onderzoek van Nicolaes Witsen (1641–1717), burgemeester en VOC-bewindhebber van Amsterdam*. Amsterdam: Bert Bakker, 2010.

Pidoplichko, I. G. "History of the Study of the Mammoth and the Beginnings of National Paleontology." *Environment and Fauna of the Past* [in Russian], no. 8 (1974): 3–10.

Pimentel. Juan. *The Rhinoceros and the Megatherium*. Cambridge, MA: Harvard University Press, 2017.

Plateri, Felicis. *Observationum in hominis affectibus plerisque corpori et animo, functionum læsione, dolore aliave molestia & vitio incommodantibus*, vol. 3. Basileæ: L. König, 1614.

Plot, Robert. *The Natural History of Oxford-Shire. Being an Essay Towards the Natural History of England*. Oxford: The Theater, 1677.

Poulakakis, Nikos, et al. "Ancient DNA forces reconsideration of evolutionary history of Mediterranean pygmy elephantids." *Biology Letters* 2 (2006) 451–454.

Rappaport, Rhoda. *When Geologists Were Historians, 1665–1750.* Ithaca, NY: Cornell University Press, 1997.

Reamur, [René Antoine Ferchault de]. "Sur les Mines de Turquoises du Royaume; sur la nature de la Matiere qu'on y trouve, & sur la maniere dont on lui donne la couleur." *Histoire de Academie Royale des Sciences* (1715): 174-202.

Richer, Jean, and Théophraste Renaudot. *Mercure françois: ou suite de l'histoire de nostre temps, sous le regne Auguste du tres-chrestien roy de France et de Navarre, Louys XIII* 3 (1614): 191–195.

Riolan, Jean. *Gigantologie histoire de la grandeure des géants.* Paris: Adrien Perier, 1618.

Riolan, Jean. *Gigantomachie pour respondre à la Gigantostologie.* Paris, 1613.

Riolan, Jean. *L'imposture descouverte des os humains supposés, et faussement attribués au Roy Theutobochus.* Paris: Pierre Ramier, 1614.

Rookmaaker, L. C., et al. "Petrus Camper's study of the Javan rhinoceros (Rhinoceros sondaicus) and its influence on Georges Cuvier." *Bijdragen tot de Dierkunde* 52, no. 2 (1982): 121–136.

Rudbeck, Olavi. "Epistola ad Virum Celeberromum Dn. Fabianum Törner de Esthonum, Fennonum Laponumque origine." *Acta literaria Sueciae* 2, no. 3 (1727): 300–306.

Rudwick, Martin J. S. *Bursting the Limits of Time: The Reconstruction of Geohistory in the Age of Revolution.* Chicago: University of Chicago Press, 2005.

Rudwick, Martin J. S. *Georges Cuvier, Fossil Bones, and Geological Catastrophes: New Translations and Interpretations of the Primary Texts.* Chicago: University of Chicago Press, 1998.

Rudwick, Martin J. S. *The Meaning of Fossils: Episodes in the History of Palaeontology.* Chicago: University of Chicago Press, 1985.

Scaramucci, Giovanni Battista. *Meditationes Familiares Circa Petrifactiones.* Urbino: Litteria Leonardi, 1697.

Schnapper, Antoine. "Persistance des géants." *Annales: Économies, Sociétés, Civilisations* 41, no. 1 (1986): 177–200.

Schrenck, Leopold von. "Bericht über neuerdings im Norden Sibirien's angeblich zum Vorschein gekommene Mammuthe." *Bulletin de l'Academie imperiale des sciences de St. Petersbourg* 16 (1871): 147–173.

Seaver, Kirsten A. "Desirable Teeth: The Medieval Trade in Arctic and African Ivory." *Journal of Global History* 4 (2009): 271–292.

Seaver, Kirsten A. "'A very common and usuall trade': The Relationship between Cartographic Perceptions and Fishing in the Davis Strait c. 1500–1550." *British Library Journal* 22, no. 1 (1996): 1–24.

Semonin, Paul. *American Monster: How the Nation's First Prehistoric Creature Became a Symbol of National Identity.* New York: New York University Press, 2000.

Shepard, Odell. *The Lore of the Unicorn.* New York: Houghton Mifflin, 1930.

Siegelt, Christine, Lutz Schirrmeister, and Olga Babiy. "The Sedimentological, Mineralogical and Geochemical Composition of Late Pleistocene Deposits from the Ice Complex on the Bykovsky Peninsula, Northern Siberia." *Polarforschung* 70 (2002): 3–11.

Simmons, John S. G. "H. W. Ludolf and the Printing of his Grammatica Russica at Oxford in 1696." *Oxford Slavonic Papers* 1 (1950): 104–119.

Simmons J. S. G., and B. O. Unbegaun. "Slavonic Manuscript Vocabularies in the Bodelian Library." *Oxford Slavonic Papers* 2 (1951): 118–127.

Simpson, George Gaylord. "The Beginnings of Vertebrate Paleontology in North America." *Proceedings, American Philosophical Society* (Part 1), 86 (1943): 130–188.

Simpson, George Gaylord. "The Discovery of Fossil Vertebrates in North America." *Journal of Paleontology* 17, no. 1 (1943): 26–38.

Sloane, Hans. "An Account of Elephants Teeth and Bones Found under Ground." *Philosophical Transactions of the Royal Society* 35 (1727–1728): 457–471.

Sloane, Hans. "Of Fossile Teeth and Bones of Elephants. Part the Second." *Philosophical Transactions of the Royal Society* 35 (1727–1728): 497–514.

Spleiss, David. *Oedipus Osteolithologicus, Seu Dissertatio Historico-Physica, De Cornibus Et Ossibus Fossilibus Canstadiensibus.* Schaffhausen: Johann Rudolph Frey, 1701.

Stanford, Donald E. "The Giant Bones of Claverack, New York, 1705." *New York History* 40, no. 1 (1959): 47–61.

Stejneger, Leonhard. *Georg Wilhelm Steller, the Pioneer of Alaskan Natural History.* Cambridge, MA: Harvard University Press, 1936.

Stellero, Georg Wilhelm. "De Bestiis Marinis." *Novi commentarii Academiae scientiarum imperialis Petropolitanae* 2 (1749): 289–366.

Stevens, Sylvester K., and Donald H. Kent, eds. *The Expedition of Baron de Longueuil,* 2nd & rev. ed. Harrisburg, PA: Erie County Historical Society, Pennsylvania Historical Commission, 1941.

Strahlenberg, Philipp Johann von. *An historico-geographical description of the north and eastern parts of Europe and Asia: but more particularily of Russia, Siberia, and Great Tartary.* London: W. Innys and R. Manby, 1738.

Strickland, Lloyd. "How Modern Was Leibniz's Biology?" *Studia Leibnitiana* 37, no. 2 (2005): 186–207.

Tamizey de Larroque, Philippe. "Un Document Inédit Sur Le Géant Theutobocus." *Bulletin du bibliophile et du bibliothécaire* (1888): 309–313.

Tamizey de Larroque, Philippe, ed. *Lettres de Peiresc.* 7 vols. Paris: Imprimerie Nationale, 1888–1898.

Tassy, Pascal. "L'émergence du concept d'espèce fossile: le mastodonte américain (Proboscidea, Mammalia) entre clarté et confusion." *Geodiversitas* 24, no. 2 (2002): 263–294.

Tassy, Pascal, and Jeheskel Shoshani. "Historical overview of the classification and phlogeny of the Proboscidea." In *The Proboscidea: Evolution and Palaeoecology of Elephants and Their Relatives.* Oxford: Oxford University Press, 1996.

Tatischow, Basil. "Of the bones of the animal, which the Russians call Mamont." *Acta Germanica: Or, the literary memoirs of Germany, &c.* (1743): 269–273.

Tatishchev, Vasily. "Tales of the Mammoth Beast, Which the Common Siberians Allege Lives Underground, with the Evidence and the Opinions of Others." *Environment and Fauna of the Past* [in Russian], no. 8 (1974): 11–28.

Tentzel, Wilhem Ernst. *Monatliche Unterredungen einiger guten Freunde von Allerhand Büchern und andern annemlichen Geschichten. Allen Liebhabern Der Curiositäten zur Ergetzligkeit und Nachsinnen.* 12 vols. Leipzig: Thomas Fritsch, 1689–1698.

Tentzelii, Wilhelmi Ernesti [Wilhem Ernst Tentzel]. *Epistola de sceleto elephantina tonnae nuper effosso, ad Antonium Magliabechium.* Gotha: 1696.

Thomas, M., et al. Supporting online material for "Whole-Genome Shotgun Sequencing of Mitochondria from Ancient Hair Shafts." *Science* 317, no. 1927 (2007).

Thompson, Keith. *Before Darwin: Reconciling God and Nature.* New Haven, CT: Yale University Press, 2005.

Tilesio [Wilhelm Gottlieb Tilesius]. "De skeleto mammonteo Sibirico ad maris glacialis littora anno 1807 effosso, cui praemissae Elephantini generis specierum distinctiones." *Mémoires de l'Académie Impériale des Sciences de St. Pétersbourg* 5 (1815): 406–514.

Timkowky, George, and Julius von Klaproth. *Travels of the Russian mission through Mongolia to China, and Residence in Peking, in the Years 1820–1821*. London: Longman, Rees, Orme, Brown, and Green, 1827.

Tolmachoff, I. P. "The Carcasses of the Mammoth and Rhinoceros Found in the Frozen Ground of Siberia." *Transactions, American Philosophical Society* 23 (1929): 1–81.

Torrubia, Joseph. "Gigantologia Española." In *Aparato Para la Historia Natural Española: Muchas dissertaciones physicas, especialmente sobre el Diluvio*. Madrid: Don Augustin de Gordejuela y Sierra, 1754.

Toulouse, Sarah. "Marine Cartography and Navigation in Renaissance France." In *The History of Cartography, Volume 3, Cartography in the European Renaissance*. Chicago: University of Chicago Press, 2007.

Turner, George. "Memoir of the Extraneous Fossils denominated Mammoth Bones; principally designed to show, that they are the remains of more than one species of nondescript Animals." *Transactions of the American Philosophical Society* 4 (1799): 510–518.

"Une Expédition Canadienne a la Louisiane en 1739–1740." *Rapport de l'archiviste de la province de Québec pour 1922–1923* (1923): 156–199.

Valentini, Michael Bernard. *Museum museorum oder vollständige Schau-Bühne aller Materialen und Specereyen nebst deren natürlichen Beschreibung, Election, Nutzen und Gebrauch*, vol. 2. Franckfurt am Mayn: Johann David Sunners, 1704.

Van Duyzer, Chet. *Legends on Martin Waldseemüller's Carta Marina of 1516*. http://www.loc.gov/today/cyberlc/feature_wdesc.php?rec=5539, accessed July 11, 2013.

Van Duzer, Chet. *Sea Monsters on Medieval and Renaissance Maps*. London: British Library, 2013.

Van Duzer, Chet. *The World for a King: Pierre Desceliers' Map of 1550*. London: British Library Publishing, 2015.

Vermeulen, Han F. *Before Boas: The Genesis of Ethnography and Ethnology in the German Enlightenment*. Lincoln: University of Nebraska Press, 2015.

Vucinich, Alexander. Science in Russian Culture: A History to 1860. Stanford: Stanford University Press, 1963.

Walkendorf, Erik. "En kort og summarisk Beskrivelse over Nidaros Diocese og særligt over den Del af denne, som kaldes Finmarken, det yderste Landskab i Kristenheden mod Nord." *Det Norske geografiske selskabs aarbok* 12 (1902): 5–23.

Walker, John Fredrick. *Ivory's Ghosts: The White Gold of History and the Fate of Elephants*. New York: Atlantic Monthly Press, 2009.

Wilkins, John S. *Species: A History of an Idea*. Berkeley: University of California Press, 2009.

Witsen, Nicolaas. *Noord en Oost Tartarye; Behelzenden verscheide byzondere gewesten, in't Noorder en Ooster Asiatische en Europische Tartarye gelegn*. Amsterdam, 1692.

Witsen, Nicolaas. *Noord en Oost Tartarye, ofte bondig ontwerp van eenige dier landen en volken, welke voormaels bekent zijn geweest*. Amsterdam: M. Schalekamp, 1785.

Witsen, Nicolaes. *Aeloude en hedendaegsche scheepsbouw en bestier*. Amsterdam: Casparus Commelijn, 1671.

Witsen, Nicolaes. *Architectura navalis et regimen nauticum*. Amsterdam: Pieter en Joan Blaeu, 1690.

Young, Davis A. *The Biblical Flood: A Case Study of the Church's Response to Extrabiblical Evidence.* Carlisle, PA: Paternoster Press, 1995.

Zárate, Agustin de. *Histoire de la découverte et de la conquête du Pérou*, vol. 1, trans. S.D.C. Amsterdam: Loius de Lorme, 1700.

MAPS

Anon. *Vallard Atlas.* 1547.

Bellin, N. *Carte de la Louisiane Cours du Mississippi et Pais Voisines.* 1744.

Desceliers, Pierre. Untitled world map. 1546.

Desceliers, Pierre. Untitled world map. 1550.

Desceliers, Pierre. Untitled world map. 1553.

de Lisle, Guillaume. *Carte de Tartarie.* Paris, 1706.

Fries, Lorenz. *Carta Marina.* 1522.

Petermann, A. *Nordpolarkarte zur Übersicht einiger geschichtlichen Momente & der jetzigen Hauptplätze der Grossfischereien (Walfischfang und Robbenschlag).* Gotha, 1869.

Waldseemüller, Martin. *Carta Marina.* Strasbourg, 1516.

Witsen, Nicolai, and Carolo Allard. *Tartaria, sive Magni Cham Imperium.* 1690.

INDEX